Advanced Sciences and Technologies for Security Applications

Indexed by SCOPUS

The series Advanced Sciences and Technologies for Security Applications comprises interdisciplinary research covering the theory, foundations and domain-specific topics pertaining to security. Publications within the series are peer-reviewed monographs and edited works in the areas of:

- biological and chemical threat recognition and detection (e.g., biosensors, aerosols, forensics)
- crisis and disaster management
- terrorism
- cyber security and secure information systems (e.g., encryption, optical and photonic systems)
- traditional and non-traditional security
- energy, food and resource security
- economic security and securitization (including associated infrastructures)
- transnational crime
- human security and health security
- social, political and psychological aspects of security
- recognition and identification (e.g., optical imaging, biometrics, authentication and verification)
- smart surveillance systems
- applications of theoretical frameworks and methodologies (e.g., grounded theory, complexity, network sciences, modelling and simulation).

Together, the high-quality contributions to this series provide a cross-disciplinary overview of forefront research endeavours aiming to make the world a safer place.

The editors encourage prospective authors to correspond with them in advance of submitting a manuscript. Submission of manuscripts should be made to the Editor-in-Chief or one of the Editors.

Manousos E. Kambouris

Preparing for the Meta-Biothreat: Disruptive Biosecurity

Adaptive Bioresilience for Responding to Emerging Biothreats

Springer

Manousos E. Kambouris
Department of Pharmacy
University of Patras
Patras, Greece

With Contrib. by
Michelle Andrews
Department of Management
J B Medical Ltd.
Bakers Mill, UK

ISSN 1613-5113 ISSN 2363-9466 (electronic)
Advanced Sciences and Technologies for Security Applications
ISBN 978-3-032-25689-8 ISBN 978-3-032-25690-4 (eBook)
https://doi.org/10.1007/978-3-032-25690-4

This Springer imprint is published by the registered company Springer Nature Switzerland AG
The registered company address is: Gewerbestrasse 11, 6330 Cham, Switzerland

To the Wilderness Feline that marked my life,
for her care
To the shadows of my sage Grandfathers,
Manousos and Dimitris, for their teachings
And to the Canadian branch of my family,
wishing them fortitude, longevity and courage

Preface I

This book expands the Biothreat iteration of the book *Hybrid Warfare 2.2*. It is prompted by the stagnancy in the cognitive level of the Biothreats at an era of explosive science and technology advancements in depth and width in biosciences, which expedite, fine-tune and *de novo* bring forth material and immaterial elements of similar or dissimilar disciplines and nature; such are the cyberoperations and micromissiles or drone technology (all brought forward by the confrontations in Ukraine and Middle East) that may increase the latent biorisk manifold. Whether this is intentional or spontaneous is irrelevant. It does consist risk amplification.

The book revolves around the fact that the biothreat is extremely diverse regarding is scope and targets and may thus be characterized as Total Biothreat, to accommodate the sum of actual and probable, or even just possible iterations and forms. A quantum leap forward, which includes biothreats without bioagents, or biothreats whence bioagents are leveraged disproportionally to their malicious potential, as in Infodemics, consists the Meta-Biothreat.

The Biothreat sensu lato appears as a stand-alone security entity or within the framework of hybrid operations and has a very complex constitution and murky prognostic markers. More explicitly, despite the actual and projected interface of the Biothreat with different aspects of technology and operations, such as engineering and informatics, it constitutes even by itself a high risk area, and a very heterogeneous one. The analysis of the different subcases by a variety of metrics is one of the basic originalities of this book.

This book ultimately is about suggesting a disruptive approach to biosecurity. And, also, about the understanding that, contrary to the vogue of using AI for everything, in here the fast turnover of non-advanced but aggressively used bioagents in surging mode may indeed be disrupted and/or derailed only by the use of AI to identify liabilities and develop or select the proper countermeasures.

The mainstream defense issues, which are approachable easily through the Internet in a basic search are used as are, with the least development. On the contrary, the bioscientific issues need some deeper knowledge to pursue and are thus more meticulously referenced.

Patras, Greece Manousos E. Kambouris

Preface II

This is a descriptive book regarding a new dimension of biothreat as it *may* materialize and a potential concept to prepare and act so as to counter it; it is speculative by definition, as there seem to be quite a number of such books detailing some or other notion of biothreat and/or bioterrorism. Speculation is slandered, and regarded as antithetic to well-tempered and substantiated scientific statement, but in its essence is not unscientific: Risk assessments, market assessments, everything is speculative when one looks into the future, and in many cases into the past. There may be no "grounded" content, to which speculative content may be compared unfavorably in the realm of projections, but educated guesses versus wild guesses. Method and substantiation always apply: in statements for the Present, judgments of the Past and projections to the Future. The analytical constituent if projected to the future, which is the main theme of this book (and projected it is in an unconventional way), becomes by definition "speculative", a characteristic allowed and occasionally sought for by respectable journals for some kinds of scientific publications/papers (as, for example, in "perspective"-type submissions in the journal Future Microbiology of the Future Medicine Group where "speculative, forward looking and even visionary" context is required[1]).

People prefer the less absurd educated guesses, but this makes the more absurd ones—often derated as "alarmist"-occasionally VERY accurate: they are scorned as wildly improbable and thus no one is prepared for such proceedings; priorities make preparedness a lame runner. If in 2005 one would say that drones would be a massive civil market, which would inevitably cause security issues once their use would be deemed "recreational", (s)he would have been stoned as "alarmist". In 2011 the ISIS used them to bomb Hmeimim airbase, in Syria, tossing light munitions attached to 3-D printed plastic tails for stability in drop. On 1 June 2025 these hit 2000 miles inside Russia by being deployed in trucks and warehouses on-site and remotely initiated and controlled. These facts speak for the validity of *alarmism* as a behavioral pattern; of *grounded* views and visions in discussing the future and,

[1] https://files.taylorandfrancis.com/expert-medicine-author-guidelines.pdf?_ga=2.64637679.194389161.1760005528-421611011.1760005527.

above all, of *peer-reviewed* scientific projections. Peer-review shows the current and immediately projected level of capability through a lens of political-correctism; it cannot predict the executive choices, nor the innovative uses, nor, actually, any event of disruptive evolution (a good benevolent example of the latter is the CRISPR method in Molecular Biology). Would anyone but deranged minds predict the use of anything like Auschwitz before WW-II? But it did happen. In the end, "Alarmist/Alarmism" is a matter of perspective: the fact that the combined nuclear arsenal in the Cold War could wipe out Life on the Planet *repeatedly* was undeniable; *living by such fear*, that a mistake might cause such havoc (a real possibility) might have been *alarmist*. At the end, alarmism prevailed and both superpowers decreased their nuclear arsenals *exactly* to avoid such eventualities, no matter how remote-and alarmist- have they been. The whole issue of Climate Change, which is now uprated in Climate Crisis, is a perfect example of Alarmism since the 70ies. But perhaps the best case of alarmism are peer-reviewed papers regarding agentless bioevents.[2] These may be considered *alarmist* as an understatement; still, no-one calls them that, nor rises ethical and legal issues. Alarmism seems to be a new trend in crowd control; alarmism against some dangers is considered Fake News and occasionally prosecuted. But against other, acceptable risks, actual, provisional or even imaginary, alarmism and crisis-mongering seems to be the newest *mantra* in crowd control. Not without precedent, historically speaking, but with renewed vigor.

This book presents in a speculative way the possible malevolent practices with a biological causative dimension. It details why and how it may happen, but not in how to do it for various and obvious reasons. The former offers credibility and consists an argument, the latter would be an open invitation to prospective perpetrators. The focus of the book is on what is needed for Bioresilience in the future. Initially, to make obvious that the evolution (both spontaneous and accelerated) and the developments in the diverse fields that formulate the Biothreat constitute (or MAY constitute) a paradigm change, which justifies the term Meta-Biothreat. Subsequently, to start thinking in different terms, conceptually, technologically and economically—in this order. The proper course of action is first to get the concept (the need for and applicability and usefulness of any tools) and only then figure out the details and peculiarities of response and preparedness. Once this is agreed upon, practicalities as how their development, use, maintenance/update, general support will be funded and by whom they shall be owned/managed will be hammered out by interested parties- and not in a uniform way, one may add.

In this context, both ethical and feasibility issues are irrelevant for the content at hand. There is nothing unethical in identifying possible malevolent uses in the days where Gain-of-Function research is considered acceptable for "studying" possible natural events to prepare for; events that may occur or not in a stochastic manner. This book describes possible malevolent uses so as to keep aware, based on a deterministic, not stochastic, basis; the malefactors may proceed by following a selection of possible courses for a variable mix of reasons. This has nothing to do with ethical issues of any kind: the bioperpetrators are usually immune to such niceties, and in a world

[2] https://doi.org/10.1089/hs.2020.0038; https://doi.org/10.1080/23779497.2020.1865182.

of diffused specialized knowledge they do not expect to obtain tutoring from any published work, at least not in a conceptual level. Nor to be prompted by it or be introduced to excuses and justifications; they are grown beyond them. As a further disclaimer, it must be stressed that no-one is prompted to follow such examples, nor told how to do so. But it is described what such pathways may be in the near future. Many issues and entities, material and immaterial, knowledge included, may be characterized as "dual-use"; Plato in the 4th century BC succinctly declared so.[3] This is once more a mental construct, and a nuanced at that, to allow sanctions in select amenities in select cases through selective ethical and legal sensitivities. Actually there are very few, if any, things that cannot be used malevolently, either by their proper use or by its denial; a clear case of the latter are the edibles, the water and healthcare/medication; and the education and the Enlightment.

There is some controversy over the actuality of "dual-use". In essence this construct is not readily applicable in things Biosecurity. At least not so since the 2001–2005 AD timeframe, when the first superviruses (called thus for lack of a better and catchy term) were manufactured with what was to be known as "Gain-of-Function"/GoF process, an euphemism for creating a possibly weaponizable definitely dangerous species/strain by genetic manipulation. The proceedings and the operative idea of the undertaking and its most worrisome results were published in peer-review journals,[4] without any fuss or outrage for both the process/research and its publication. This was far from Responsible Publishing by authors and journal/publisher, although it was occurring after the (moderately lethal) Anthrax Attacks of 2001 in the US.[5]

Additionally, methodology allowing viral genomes to be produced from scratch from commercially available parts was also published,[6] once more no objections raised. Detailed publications regarding the threat of the Black Biology—the sinister part of the kitchen or DIY- Biology movement[7] also surfaced, involuntarily instigating relevant interest primed by curiosity or malice.[8] By extending an argument set above, it is doubtful that there is even one biomedical technology that cannot be used malevolently, *especially* Synthetic Biology.[9] The threat projection is not, by its nature, "alarmist", ethical or legal. The knife is a tool, with a given set of possible applications; these are actual qualities. How to use it is a choice, and this choice is an ethical and legal issue. Describing how probable and destructive such use may

[3] "Science devoid of Justice and other virtue is wickedness, not wisdom" Menex (XIX 247 A).

[4] https://doi.org/10.1128/jvi.75.3.1205-1210.2001.

[5] https://doi.org/10.1001/jama.286.20.2554.

[6] https://doi.org/10.1073/pnas.2237126100.

[7] https://doi.org/10.1038/embor.2009.145.

[8] https://doi.org/10.1093/embo-reports/kvd083.

[9] DiEuliis D 2019. "Key National Security Questions for the Future of Synthetic Biology." The Fletcher Forum of World Affairs 43(1): 127–43. http://www.jstor.org/stable/45289832.; https://www.homelandsecuritynewswire.com/dr20170901-biosecurity-and-synthetic-biology-it-is-time-to-get-serious.

be, a discussion which may lead proactively to better wound care and legal regulation is not ethically debatable, nor legally ambivalent, and definitely not touching on dual-use. There is no misinterpretation or controversy in here. Either a product of intelligence is produced, aimed to advance our horizon, or not. Malefactors (both non-state and state- sponsored ones) will have their own horizons extended with or without such products; they are evil (and even this might be questionable), not stupid (this is unquestionable).

Furthermore, the innovative core of this book consists of the different aspects of biosecurity as causally interacting and interrelating with actual and emerging defense/security issues, such as missile-centric warfare, the explosive development of drones of every kind and size and the increased dependence on IT amenities for exploiting the above; issues up to now examined in a most passing and casual way as dimensions within the original and innovative such concepts. The book Hybrid Warfare 2.2 of 2023 does exactly that, by considering the Biothreat as an integral part of the hybrid threats scope. In this title, the evolving, projected Biothreat and the innovative pathways (Disruptive Biosecurity) to tackle it, as an integrated antithetic pair, is the protagonist and other interacting entities are mentioned as peripheral; it is a difference of aspect and thus of detail. As a result, the focus is on the assets, liabilities, synergies and incompatibilities of the different sectors expected to define the hybrid threat from now on. To fully appreciate the extent and depth of the work presupposes audience familiar with security and operations ideas but not expert in bioscience nor in any field of engineering or political/social sciences.

Patras, Greece Manousos E. Kambouris

Acknowledgements The author wishes to express his gratitude to the following colleagues who assisted in the publications on which some of the Chapters have been based in terms of ideas and scientific theses:

Yiannis Manoussopoulos

ELGO-Demeter, Plant Protection Division of Patras, Patras, Greece
https://orcid.org/0000-0002-6065-7368, inminz@gmail.com

George P. Patrinos

- Department of Pharmacy, University of Patras, Patras, Greece
- Hellenic Pasteur Institute, Athens, Greece
- Department of Genetics and Genomics, College of Medicine and Health Sciences, Zayed Center for Health Sciences, United Arab Emirates University, Al-Ain, Abu Dhabi, UAE

https://orcid.org/0000-0002-0519-7776, gpatrinos@upatras.gr

Aristea Velegraki

- Department of Microbiology, Medical School, National and Kapodistrian University of Athens
- Mycology Laboratory, BIOIATRIKE SA, Athens, Greece

https://orcid.org/0000-0002-7605-0210, aveleg@med.uoa.gr

Competing Interests The authors have no competing interests to declare that are relevant to the content of this manuscript.

Contents

Part I
Current, Evolving and Emerging Actuality

Chapter 1
Introduction and Definitions

Abstract The perpetrated biothreat is intermittently present in different public fora. In a quest for normality, it is assigned lower priority. The Total Biothreat is considerably expanded, consisting of biothreats spontaneous or perpetrated, rudimentary or advanced, simple or complex and single- or multi-agent, possibly interacting spatiotemporally and/or functionally. It is a concept proposed to accommodate the sum of actual and probable, or even just possible iterations and forms such as Biocrime, Bioterrorism, Biowarfare, Agroterrorism and Agrocrime and Ecoterrorism and Ecocrime. If biothreats without bioagents-"agentless", or biothreats whence bioagents are leveraged disproportionally through other factors, such as hybrid, cyber or psychological operations, are factored in, the Meta-Biothreat emerges. Regarding the nature and manipulation/processing of the agents use, there may be 3 levels of biothreat: Cutting edge, Mainstream and Ubiquitous; all three may be both destructive and disruptive, but to different degrees. The "Total Biosecurity" concept, is aimed against the Meta-Biothreat. "Disruptive Biosecurity" is an iteration towards a more adaptable Modus Operanti, with focus on the liabilities of the entire attack/perpetration process, from inception to impact. It proposes not blunting, smashing or taking away the weapon, but unoptimizing every step of the design, execution and leveraging or subsequent exploitation of the perpetration event.

Keywords Agroterrorism · Ecoterrorism · Biowarfare · Biocrime · Bioterrorism · Hybrid operations · Meta-Biothreat · Disruptive biosecurity

The issue of perpetrated biothreats as a major risk factor has an intermittent presence in the different levels of the public debate and a similarly unstable position in the priority lists of decision-makers of the public sector/governance establishments—the latter understood as the diverse levels of government officials plus international organizations, sponsored or constituted from sovereign states. The biothreat should be apperceived as inherently perpetrated, which suits better the literary overtone of the "threat" root, while the biorisk may be of any and unknown, or undefined causality. Alternatively, a two-step process may be understood, where a Risk evolves to a Threat, which, if not contained or resolved, will materialize to Event. Thus the

M. E. Kambouris, *Preparing for the Meta-Biothreat: Disruptive Biosecurity*, Advanced Sciences and Technologies for Security Applications,
https://doi.org/10.1007/978-3-032-25690-4_1

prospect of diverse biothreats of any magnitude is treated as being inherently different than the—sudden—apprehension over spontaneous outbreaks of epidemic nature. It is true that the spontaneity of the latter *cannot* be *solidly proven* but is attested by lack of proof for the opposite, i.e. for a perpetrated causality. In a quest for normality, the biothreat is assigned lower priority in order to avoid any excess panic due to sentimental focus by the public, as the abstract natural forces and possible collective suffering and abuse, exemplified by the Climate Change/Crisis *mantra* [1] are far easier to manipulate during the crisis management, and long after such crisis may have been resolved. The result is that the perpetrated biothreat comes and goes from the public sphere, and interest in it is intermittent if not periodical. The reasons are not part of this discussion; but the tangible results are.

The perpetrated biothreat is squeezed as an eventuality somewhere between security threats, tackled by such services, lately called "Law enforcement community", and what has been termed Public Health Emergency -of International Concern (PHEIC)- or not. As evident by the COVID-19 pandemic, the exemplary PHEIC as declared by WHO [2], the consensus seems to be to summarily delegate such issues to a more cohesive and internationally minded management community, the medical and public health sector. The former is a part of the latter, although the more abundantly financed and celebrated part. Whether this is a good, or even a functional idea, is debatable: some countries, as were China and Russia, mobilized assets and protocols earmarked for bioterrorism/biowarfare [3]. The grim efficiency of such options plus their suitability should a perpetrator is involved might be a more efficient and effective choice.

1.1 The Total Biothreat

The Total Biothreat engulfs a much expanded and continuously evolving threat range consisting of biothreats that may actually be spontaneous or perpetrated, rudimentary or advanced, simple or complex and of one or more agents of different pedigrees, possibly interacting spatiotemporally and/or functionally [4, 5].

1.1.1 Different Kinds and Types of the Biothreat. The "3Bs"

The Biothreat is extremely diverse; it has always been so, but nowadays even more, and, which is more important, for a variety of reasons, currently this fact is understood, accepted and included in relative discussions and pondering. Based on scale, but also on a range of other markers, the first division of the Biothreat may be among Biowarfare, Bioterrorism and Biocrime, the "3Bs". These are parts of the all-encompassing biothreat of intentional human agency but differ first and foremost in Scale, as already mentioned. Biocrime relates to individual or very small scale [6], Biowarfare to massive and Bioterrorism to anything in between and including both [7]. All the above are of course subject to interpretation. Perpetration is another marker and relates to the (bio)offender's status, basically a three-category division again: the perpetrators may be state agents proper (once more there is much discussion in what may qualify as a "state", especially for legal and regulatory purposes in this particular case); they may be state-sponsored proxies, within the vast realm of irregular and hybrid operations, which makes the field very heterogeneous but rather easier to define. And then there are the non-state actors, where there is chaos, and although it seems very easy to define, hybrid operations for achieving deniability make once more the recognition and qualification of this category a very perplexed issue. For example Hamas and, long before it, the Palestinian PLO, the Irish IRA, the Armenian Asala, the Tamil Tigers and the Jewish Stern and Haganah were non-state actors, as they were attempting to secure a state in the future, but were/are attached to the territory of the prospective state and its population and the diverse characteristics of the latter: ethics, religion, history, and culture. The Hezbollah and the Kosovo Liberation Army were/are state-sponsored; perhaps the former is one of the clearest and best examples, along with the Afghan Mujahedeen of the 80s. But the six previous ones were affiliated with a State-to-Be. And then there is the ISIS-K and the Antifa cases. These do have their seats of power somewhere, but, for the moment, they are not State-sponsored; there seems to be private funding to quite a degree and then self-funding from activities, mainly but not entirely illicit. No state funding seems to be present, at least not reminiscent of the direct sponsorship of the Afghan Mujahedeen and Hezbollah (Stern, Asala, Tamil Tigers, PLO and Haganah were considerably different cases).

The next iteration is the Objectives pursued by the perpetrators [7] and, finally, the available or aspired executive Means. Both markers are very diverse to analyze and utterly linked to interpretation by one regulatory or taxonomic system or other and the respective definitions. Crime usually links to monetary gains and immunity from any prosecution, but, in this respect, in order to destroy competition and secure respect/recognition/prominence, the policies of certain states, for example the Opium Wars of the British Empire and recent proceedings by the US with the assistance of a completely docile and ill-defined "international community", raise an eyebrow. Thus did, of course, the very similar policies of Republican and Imperial Rome, Achaemenid Persia and of the Athenian Empire. In all such cases, the language is irritatingly similar regarding securing income (sanctions-international currency-favored/privileged status in mutual commercial ties), precedence (the duty/right to lead) and lack of any kind of accountability due to a mandate, either from Fortune, Virtue and Fortitude (Rome), or from some divine decree (USA, Achaemenid Empire) or previous own achievements (Athenian Empire).

Additionally to the above complications, there are more internal and external factors that perplex a straightforward assignment of a case to the three abovementioned categories: Special Operations and even more so the conduct of Secret and Hybrid Warfare and the proxy wars blur any distinction based on the Perpetrators' nature. At the same time, the interface of Biocrime with Bioterrorism (the current consideration of Drug Cartels, a flatly criminal category of organizations, as of Terrorist status by the USA) and of Biowarfare with Bioterrorism (irregular fighters are characterized legal freedom fighters by own sympathizers and terrorists by the enemy, ample examples being the Afghan Mujahedeen, the Greek-Cypriot EOKA, the Jewish Stern, the Palestinian PLO, and the two British active-duty army officers Blake and Caroll, caught by Russian troops to cooperate with the Ukrainian armed forces in Ochakov in 2025) compromise the validity of the two markers "Objective" and "Means". Moreover, the Hybrid Warfare by itself and the inherently cascading nature of some biothreats both put the "Scale" prerogative in doubt, as seen with the Black Death: meant to be a local incident by the Mongols who were besieging Caffa, it spread and devastated the whole of Europe for decades [8].

Regarding the nature and the degree of manipulation/processing of the agents to be used, there may be 3 levels of biothreat: (1) Extra high-tech/cutting edge; (2) Mainstream/state-of-the-art; and (3) Ubiquitous, widely accessible options. All may be used in both destructive and disruptive contexts, but to different degrees and with different optima.

The High-tech iteration comprises (i) xenobiology, astrobiology/exobiology (as with "space bacteria"); (ii) conventional synthetic biology, including de-extinction efforts and bionics-hybrid systems, and, of course, (iii) the agentless events; all with or without an AI dimension. The above are *revolutionary* iterations which may prove highly disruptive just by the impression their (intent of) use would make.

The Mainstream iteration is a conventional capability for well-funded and provided for perpetrators, with destructive effects variable in level and intended impact (attrition/overthrow). The Gain-of-Function endeavors, where new (almost always detrimental/malicious) attributes are added to a given agent, mostly (but not necessarily exclusively) through genetic manipulation [9–11], lie in the heart of this iteration [12] and constitute the manipulated agents category, although there are other cases. There is of course the category of the entirely artificial agents, assembled and designed from scratch or from simple bits and pieces, including minimalistic cell chassis for pasting full or artificially assembled genomes [13].

The Ubiquitous iteration focuses on accessible, affordable bioagents; "affordable" in terms of acquisition/production in massive scale and repeated batches, in centralized or dispersed but interrelated and coordinated formats—the latter being an often overlooked issue. Its massiveness, unpredictability and adaptability may be disproportionally disruptive, by the surprise it entails and the unsettling of established SOPs, routines and concepts, which are usually attuned to centralized practices and/or high tech iterations. Thus a similar, multiple, successive and/or dispersed deployment of obsolete and even moderately effective agents may smash confidence and morale amongst the targeted population/entity, irrespective of the exact health casualties.

If used in relatively small scale, and more or less locally, by causing even a minimal destructive effect, such agents would instigate any response system to react in a standard, integrated, centralized manner, as in the case of COVID-19 [14] or-preferably—better and more swiftly. By the time countermeasures would be devised and deployed, even in a superfluous manner, to bolster the confidence of the affected communities and individuals to the authorities, with some (considerable) financial and social costs already incurred, another, similar agent may emerge at the same places or elsewhere, to draw resources, attention and intervention anew. Thus, either the lockdowns and other restrictive measures will be renewed to already impacted or /and expanded to new localities; or further undermining of the morale will ensue, causing a surging effect in an already affected locality/population. This leads to the saturation of the responses which are and will be optimized for a "one-at-a-time" bioevent (a "Singularity" [15]). The surge of many different, affordable and quickly replaceable agents, dispersed with minimal resources, is meant to trigger reaction (a forced move operationally but also politically and psychologically); not to evade or outsmart the reaction by the responding services. The aim is/would be to exhaust them and disrupt the targeted entity (military unit/establishment, population, polity) without the expenses and risks of truly lethal and/or advanced bioagents.

The Ubiquitous iteration may be opted for by many a prospective perpetrator. It is the analogue of the Black Biology of the 90s. Though, as a procedural choice it may occasionally attract major actors. It is an attractive, flexible option: High- or low-tech, deployable by standard operatives with no expertise and little training. It may be well-combined with drones and other, affordable and readily accessible modalities of distributed delivery, produced in small-scale setups (in-house), intended for fast and limited spatiotemporal use, to be succeeded by a next iteration, once Medical Counter-Measures/MCM and other interventions render it ineffective. It may become, as per the above, a morphic threat, continuously changing in identity and hopping in spatiotemporal context, with some destructive potential but huge disruptive one: the intended targets (people, livestock, environment, crops) keep getting sick and the surge of syndromes not only negates or perplexes therapy/care but leads to continuous application of strenuous measures such as lockdowns, quarantines, transportation restrictions etc., destroying social and organizational functions.

1.1.2 Agro- and Eco-Threats

If examined from the viewpoint of the intended targets, the Biothreat, which is usually taken in the human context, has two offshoots: Agroterrorism and Ecoterrorism. The former as a term appeared in the 1990s [16], to describe a threat which was growing as a divergent discipline of bioterrorism back in the day [17], although the term was not widely used. In late 2010s the Bioterrorism notion was attributable to plants also [6], a clear overlooking of the Agroterrorism concept. The different trade wars and the issues of international drug trafficking and commerce (re)introduced the term within Hybrid Warfare. Less disliked than harming human subjects, and less guarded against, with possibly great repercussions and asymmetrical impact as to the effort and risk of the perpetrator [18, 19], Agroterrorism seemed a very promising offence method.

The evolving and specialized efforts for countering Agroterrorism consist the field of Agrosecurity. For the generations grown in the western world, in thriving societies with an abundance of food supply (though NOT relevant production usually), Agrosecurity was a rather marginal issue. But given that, at least since 2008 [20], the specter of famine has re-emerged, Agrosecurity is no more a marginal issue referring to the income and welfare of farmers, but a vital issue of communities for securing sufficiency of vital supplies.

Agrosecurity and Agroterrorism are not *subcases* of biosecurity and bioterrorism, respectively, nor of biocrime, by any interpretation. The latter is correctly exempt, as the agrothreat always has a massive impact. Although 'crime', 'terrorism' and 'warfare' are terms used as policy dictates and largely interchangeably, the notion of biocrime usually follows that of conventional crime, meaning small-scale events. Little discussion refers to agrowarfare or agricultural biowarfare [21–24], despite its historicity: the Allies were planning to starve Germany in WW II, with means including but not limited to "Operation Vegetarian" [25, 26]; the Nazis implemented such measures against the Soviets, such as the "Hunger Plan" [27], and the Soviets had done the same against the Ukrainians with the "Holodomor" [28]. The US did so in the Sioux wars, by slaughtering the buffaloes [29], the main food source of the tribes, and in the Korean War by the mass destruction of fishery nets by the US Underwater Demolition Teams in Korea so as to starve the pro-Communist population [30]. Famines did ensue in both the cases. The issue became focal in the 1990s as the US was contemplating (and perhaps did use) bioagents to destroy crops of marijuana plants in Latin America [31].

Regarding the association of the subcases of the *agrothreats* and the *biothreats*, agroterrorism may be implemented by other, non-biological agents, and is thus larger in scope than bioterrorism. Actually, it is the terrorism *against* the *agro*-sector [32] as opposed to the terrorism *using bioagents*. Consequently, agrosecurity responds to threats *targeting* the agro sector, while biosecurity responds to threats *posed* by bioagents. As a result, a microorganism or a metabolite that is not poisonous for lifeforms but degrades a technical amenity [33, 34], such as cables (as in wired telecommunications or electric apparatuses) or rubber (as in tyres) or wood [34], is a biothreat and an issue of biosecurity. Conversely, the core of agrosecurity includes the use of inorganic chemicals to spike soil in order to make it infertile [35], or the use of radioactive waste (from nuclear plants, from weapons-grade enrichment or even from mining sites or misdisposed such waste of industrial or medical facilities) to contaminate supplies by insertion in the food chain. Thus entire stretches of land become barren and/or actual health hazards. It also includes the use of synthetic pesticides to destroy a specific crop and even the use of incendiaries to destroy crops; in both cases no bioagent is involved. Biofuel and, at a later point, bioelectricity are among the amenities that may become mainstream and essential and thus become targeted by Agrothreats. Agrosecurity and biosecurity have important common elements, but neither is a part of the other.

The issue of Ecoterrorism is very similar, if not understood in the sense of the ideological terrorism exercised to uphold and promote alleged environmental issues, as happens regularly [32]. If considered as an equivalent to Agroterrorism aimed at Wildlife and the Environment in general [36], as done in this book, there is no concept of Ecocrime; Ecowarfare sounds weird, although the US use of defoliate agents to destroy the leaf canopy and thus the jungle in Vietnam [19, 37] accounts for exactly this, and obviously may be enacted by non-biological agents. The concept of Ecowar also qualifies for a more objective description of the US slaughter of the buffalos [29,

38] as the extermination of the species changed radically the environment; possibly a most welcomed, if not intended, side-effect of the campaign, to use the plains for agriculture instead of fodder for roaming herd.

1.2 Stepping Up in the 2010s: The Meta-Biothreat

A number of reasons led to the differentiation of the Biothreat in early 2010s to many, different and advanced [39] or crude iterations [40]. Such reasons ranged from the pressure by the enhancement of the security establishments due to the renewed interest in the wake of the twenty-first century, and the preparedness efforts caused by the string of the viral epidemics worldwide, to the technological advancements that allowed various multi-disciplinary biothreats, as indicated by the intense deliberations on the subject of GCBR [41–45]. The explosive progress in drone technology and the innovative use of them, as first witnessed by ISIS in Syria at the same time [46], along with the quantum advances in informatics, computer science, as with the Internet-of-Things/IoT [47], and, ultimately, Big Data [48] and Artificial Intelligence [49, 50] created a very different landscape in the realm of the Biothreat. Informed by innovative thinking in operations, tactics and practices, the whole became much more threatening and consists what is herein termed the Meta-Biothreat. It may be understood as comprising three main sub-realms:

1.2.1 Dispersed, Surging, Fast-Track Bioattacks

The first constituent of the Meta-Biothreat is the potentially very different, actually expanded, operational concept; in a nutshell, the importance is not in what one uses (a normal choice would be Ubiquitous agents), but in how, where, when and how much of the selected asset(s) is used. Even the same items and resources are to be used in very different ways, to achieve surprise and circumvent standard response planning. Bioagents may not need to be advanced to outsmart Bioresilience efforts. Suffice to be smartly employed, for example existing Bioagents, which would be threatening and seemingly or genuinely uncommon, to be harvested and processed so as to be easily and promptly delivered in serial or surging order. Intended for local deployment, by causing a minimal destructive effect, this series of events would incentivize any response system to react in a standard, integrated, centralized manner, as in COVID-19 [34]. By the time countermeasures would be devised and deployed, to considerable financial and social costs, another, similar agent may emerge elsewhere to draw resources, attention and intervention. This almost random, quick succession, fed by the ease to readily modify existing microbial genomes and produce limited quantities of crudely weaponized, or even unprocessed agents, is the opposite of high-tech pathogen manufacture, leads to many different, affordable and quickly replaceable and replenishable offensive means dispersed with minimal resources and playing onto the need of a system to react. It is not intended to escape or outsmart reaction by the intervening services, but simply to overload, exhaust and overwhelm them and disrupt the targeted entity (military unit/establishment, population, polity).

The widespread deployment may require as a countermeasure the inclusion of specialists of some kind to very low echelons of military units or security intervention and civilian first response teams. A biospecialist may become similar to the FPV (First Person View) drone operator, the grenadier or the radio operator in an infantry squad, with a diversified but basic outfit. This may include first a portable kit for bacterial/cell transformation (a principle used in Soviet BW warheads designed for ballistic missiles), to turn local microbiota or microbiomes to more aggressive, pathogenic mutants, and/or bolt-on warheads containing pathogens, toxins or decon/sanitizing agents for attachment to hand-grenades, mortar rounds and other types of ammunition, including multipurpose/antitank portable rocket launchers/recoilless rifles for tactical deployment in active firefights.

1.2.2 Agentlessness/Cyber Leverage

The second constituent of the Meta-Biothreat is the advent of a range of options of minimal bioperpetration, where the offender does not use such amenities. These events are engineered mainly, but not exclusively, through the cyber realm [51]. The most advanced and minimalistic case refers to no bioagents being employed ("agentless event"). A concerted disinformation and misinformation campaign persuades a target population that an event is occurring, thus causing panic, mass loss of confidence to the authorities (if the event is denied) and possibly chaos, with pharmacy runs, self-prescribed or fake-news prompted prophylaxis which may cause actual reverse reactions of variable severity, plus the full range of disruptive effects, with social disorder, economic pause, and quarantines/lockdowns if the authorities are pulled into the Invented Reality.

Other options refer to the leveraging, with the same mechanism as above, of minor events to PHEIC or GCBE/ELE—Global Catastrophic Biological Event/Extinction-Level Event [6, 34, 41, 52]. There is an event, spontaneous or perpetrated (zero/minimal perpetration respectively) which may be verified, but its impact, severity, expansion and, most of all, dynamics are exaggerated intentionally. The results are similar to the abovementioned ones for the agentless event. And then, there are the indirectly perpetrated events. These are usually to occur through digital gateways, in a cyberoperation. There are widely different possibilities, from the changing the prescriptions through electronic prescription systems to unsuitable or outright unsafe drugs, to the compromising of test instruments or life support apparatuses, all the way to causing failure to safety and security features, instruments and procedures in biolabs, hospitals and other units employing pathogens of interest for research or biotechnological purposes or disposing of them as biowaste. The issue will be examined in detail, along with other dimensions of the AI in biothreats, in Chap. 5.

1.2.3 Advanced Bioagents and Delivery Methods

The third and most easily recognizable constituent of the Total Biothreat is the evolution of purely technical means, along with the possible scale and objectives and the convolution of all the above. The scale (3Bs) and objectives (Agro/Eco, technical and human targets) were discussed above. The main focus, and deservedly so, lies with the agents proper, although understood *sensu lato*, in order to include the vectoring methods, strategies and amenities and any immaterial (such as intelligence, skills, knowledge, know-how [53]) and material elements that are required to develop them; these are examined in Chaps. 3 and 4. Briefly, the selection of natural pathogens (or "agents of interest") which was the initial approach to Bioassault, implies agents with a number of qualities, the exact mix and balance of which depend upon the envisaged use and accessibility. Such qualities were (indicatively and not exhaustively) virulence, environmental stability, robustness to withstand possible shocks and spikes during deployment, compatibility with expedient and short production times and low costs, plus long conservation ("self life") without special requirements (vacuum, refrigeration) in order to be clandestinely inserted/smuggled without need for tell-tale devices and equipment, the ability to withstand sterilization and, especially, antibiotics and antiseptics.

Different technological state-of-the art, and more importantly different knowledge and cognitive basis, allow classification of the agents in generations [34]. As the manipulation becomes more radical the generation indicator increases, and this has been achieved mainly but not exclusively through genetic engineering. A further step, incorporating the more advanced iterations, increases the variability manifold: there are the fully engineered but of conventional biochemistry/molecular biology microbiota, which may be very drastically (re)designed with the help of synthetic biology [54]. More advanced are the techno-biohybrids with nano-material manufactured special, novel or similar organelles to the natural ones [55]. The most exotic are the xenobiota in all different meanings and definitions. The term may refer to extraterrestrial microorganisms of same, similar but, more obviously, of completely different biochemistry [56, 57], or to microorganisms in secluded enclaves within the planet, evolving cut off from the actual biosphere, or to inhabitants of extreme environments, human-made space habitats included and, last but not least, the synthetic biology products designed and manufactured to have minimal or rather null genetic interaction with the biosphere genomes [58]. One should include in the "exotic" category the re-emerging/de-extinct- and possibly subsequently manipulated- species and strains, as are the viral strains of the Spanish Flu [9, 53, 57, 59]. If one considers the above, the generations of microorganisms available as bioagents reaches to seven, i.e. 0–6th [34]. This is without taking into consideration the de-extinction dimension, which may or may not be included into the xenobiota, as will be discussed in Chap. 3.

It is clearly understood that the biochemical warfare, which utilizes not cells or organisms, but molecules of biological origin, such as bioregulators and, most of all, biotoxins, is counted within the biothreat realm and counts six generations, 0–5th [34]; the use of organisms, strictly speaking, defines the microbiological warfare/

threat. It is rarely understood that the biothreat, regarding live agents, extends far beyond, or rather above, microbiota. It is not just the parasites, which may be clearly visible and very efficient in causing a diverse set of symptoms and massiveness of impact [60], with the dimension of plant parasites being of renewed interest [61]. Macroorganisms proper may be used in warfare, as they satisfy a very basic and workable definition of the biothreat as used herein: "...the use of live organisms or their products and processes for criminal/terrorist/warfare purposes". The range is great: one may start with originals, like moats with crocodiles, war dogs and elephants for directly attacking enemy materiel, positions and personnel, or even cavalry mounts for transportation of assault elements and charging the enemy outright. Then there are the beasts of burden, for moving supplies and means/logistics [34] and an infinite range of harmful or disruptive organisms: insects in diverse roles and assignments [62], and pots of snakes. Soon to appear in a neighborhood near us, are the prospectively neurohacked company, wildlife and stray animals, including the guard ones (mostly, but not exclusively, dogs) guided to assume different missions by altering their behavior and the trained patrol dolphins or sharks. The macrobiota count in six generations (0–5th) as well [34].

The delivery of bioagents to their intended targets, in functional or spatiotemporal terms, is a highly diverse, multidisciplinary issue. It is seldom understood that delivery, also mentioned as Vectoring, is almost as important as the agent itself. Although in most cases entirely different than the agent, the delivery means in essence is one part of a binary, two-part system: the vector-effector dionym [4]. A given agent with different delivery/vectoring approaches might show extreme variability in efficiency: from a pandemic to a null release, as understood by the Imperial Japanese Army Unit 731 in the Chino-Japanese War in the 1930s and 1940s, where the same pathogen dispersed by bombs caused null damage, but if the bombs released insects vectoring (possibly the same strain of) the same pathogen the results were devastating [63]. At present, delivery may be enacted by a great variety of dedicated or improvised devices, of high or low tech, from pointed and sharp tainted instruments propelled by physical force, pneumatic, hydraulic or chemical energy to dispersion of airborne forms, tainting of surfaces, poisoning of supplies, malicious use of drug delivery systems and multi-stage vectoring systems comprising any or both fomites and hosts [4].

The emerging liabilities of the current concepts of Biosecurity and their implications in preparedness, prevention and response may be partly alleviated through novel commodities available for medical use, both diagnostics and therapeutics. Given the nature of the projected biothreats and their possible perpetrators, different approaches than the usual epidemiological ones should be explored, focusing on the weak or choke points of the implementation pipeline of a perpetrated event, and not on epidemiological projections based on some constant or other. This new focus may be achievable through the use of AI, since very large amounts of differently structured, curated and obtained data need to be sorted, filtered and mined to make sense.

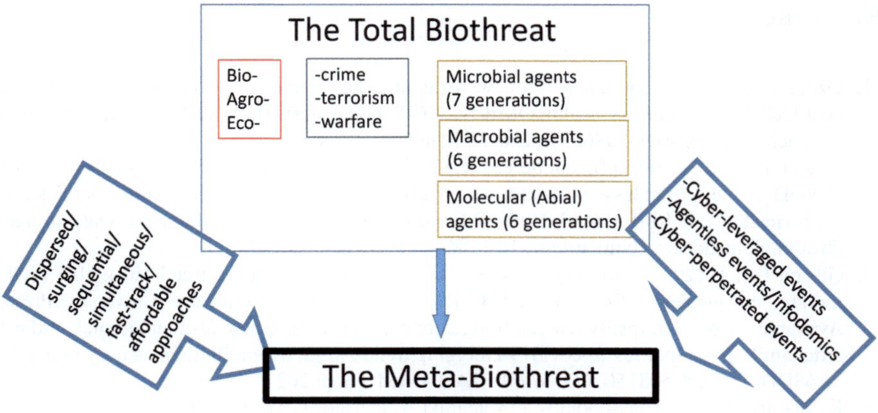

Fig. 1.1 The makeup of the Meta-Biothreat

1.3 Disruptive Biosecurity Against the Total Biothreat

The Total Biothreat is a concept proposed to accommodate the sum of actual and probable, or even just possible iterations and forms of actual bioagents, molecular or organismic in nature [5]. A quantum leap forward consists the Meta-Biothreat [4], which includes biothreats without bioagents—"agentless", or biothreats whence bioagents are leveraged disproportionally to their own malicious potential, as in Infodemics [64], along with more innovative employment of conventional, proper bioagents (irrespective if these are unmodified, crudely or elaborately processed, or fully engineered—the focus is *not* on the bioagent) (Fig. 1.1).

In this light, it could be advantageous to pursue a new biosecurity concept, which includes a novel level—and philosophy—of intervention. It is directed against the Meta-Biothreat. The idea is to combine the existing knowledge, means, and even procedures, which, conditionally, might be adequate for tackling Total Biothreat events, in a proactive and reactive concept termed herein "Disruptive Biosecurity" intended to focus on the liabilities of the entire attack/perpetration process, from inception to impact [4]. It proposes not blunting, smashing or taking away the weapon, but unoptimizing every step of the design, testing, production, storing, deployment, use and leverage/exploitation of the drastic effect, direct or indirect. This approach differs from previous ones that focus either on the management of the consequences and casualties (Reactive Biosecurity) or on interventions before a perpetrated bioevent materializes (Proactive Biosecurity); the Disruptive Biosecurity is both proactive and reactive, employing one or more of a panel of methods, means and processes adaptively to any specific risk factor before it morphs into a threat, while being a clear and present/live threat and after the threat has materialized to event/crisis.

References

1. Özdemir V (2020) "One nature": a new vocabulary and frame for governance innovation in post-COVID-19 planetary health. OMICS 24:645–648 [Internet]. Available from: https://pubmed.ncbi.nlm.nih.gov/32986539/. Cited 6 Sept 2021
2. Assefa Y, Gilks CF, Reid S, van de Pas R, Gete DG, Van Damme W (2022) Analysis of the COVID-19 pandemic: lessons towards a more effective response to public health emergencies. Glob Health 18:1–13 [Internet]. BioMed Central Ltd. Available from: https://globalizationandhealth.biomedcentral.com/articles/10.1186/s12992-022-00805-9
3. Gibson-Fall F (2021). Military responses to COVID-19, emerging trends in global civil-military engagements. Rev Int Stud 47:155–170 [Internet]. Cambridge University Press. Available from: https://www.cambridge.org/core/journals/review-of-international-studies/article/military-responses-to-covid19-emerging-trends-in-global-civilmilitary-engagements/1A68DD0D2F22C5081507FD15743FF2AD. Cited 7 Sept 2023
4. Kambouris ME, Manoussopoulos Y, Velegraki A, Patrinos GP (2025) Disruptive biosecurity: a new concept for preparedness and response to the emerging total biothreat landscape. Comp Strateg 1–33 [Internet]. Routledge. Available from: https://doi.org/10.1080/01495933.2025.2565586
5. Kambouris ME (2024) Integrating the two: an operational and strategic aspect. In: Kambouris ME (ed) Hybrid warfare 2.2: where biothreats meet irregular operations and cyber warriors in the 21st century, 1st edn. Springer Nature, Cham, pp 123–139
6. Millett P, Snyder-Beattie A (2017) Existential risk and cost-effective biosecurity. Health Secur 15:373–383
7. Jansen HJ, Breeveld FJ, Stijnis C, Grobusch MP (2014) Biological warfare, bioterrorism, and biocrime. Clin Microbiol Infect 20:488–496
8. Wheelis M (2002) Biological warfare at the 1346 siege of Caffa. Emerg Infect Dis 971–975 [Internet]. Available from: https://pubmed.ncbi.nlm.nih.gov/12194776/. Cited 28 Sept 2020
9. Tumpey T, Basler C, Aguilar P (2005) Characterization of the reconstructed 1918 Spanish influenza pandemic virus. Science 310:77–80
10. Jackson RJ, Ramsay AJ, Christensen CD, Beaton S, Hall DF, Ramshaw IA (2001) Expression of mouse interleukin-4 by a recombinant ectromelia virus suppresses cytolytic lymphocyte responses and overcomes genetic resistance to mousepox. J Virol 75:1205–1210 [Internet]. American Society for Microbiology. Available from: https://pubmed.ncbi.nlm.nih.gov/11152493/. Cited 21 Sept 2020
11. Dance A (2021) The shifting sands of "gain-of-function" research. Nature 598:554–557
12. Kaiser J (2014) Scientists call for limit on creating dangerous pathogens. Science [Internet]. Available from: https://www.science.org/content/article/scientists-call-limit-creating-dangerous-pathogens
13. Kambouris ME, Manoussopoulos Y, Kantzanou M, Velegraki A, Gaitanis G, Arabatzis M et al (2018) Rebooting bioresilience: a multi-OMICS approach to tackle global catastrophic biological risks and next-generation biothreats. Omics J Integr Biol 22:35–51
14. Chinta S, Rodriguez-Guerra M, Shaban M, Pandey N, Jaquez-Duran M, Vittorio TJ (2023) COVID-19 therapy and vaccination: a clinical narrative review. Drugs Context 12
15. Kambouris ME, Gaitanis G, Manoussopoulos Y, Arabatzis M, Kantzanou M, Kostis GD et al (2018) Humanome versus microbiome: games of dominance and pan-biosurveillance in the Omics universe. Omics J Integr Biol 22:528–538
16. DeOtte RE (2007) Agrosecurity for concentrated animal feeding operations (CAFOs): commentary on recent planning activities. Anim Health Res Rev 8:89–103 [Internet]. Available from: https://pubmed.ncbi.nlm.nih.gov/17692146/. Cited 23 Aug 2022
17. Gyles C (2010) Agroterrorism. Can Vet J 51:347–348 [Internet]. Available from: https://pubmed.ncbi.nlm.nih.gov/20592822/. Cited 23 Aug 2022
18. Khalil AT, Shinwari ZK (2015) Threats of agricultural bioterrorism to an agro dependent economy; what should be done? J Bioterror Biodef 05:1–8 [Internet]. OMICS Publishing

Group. Available from: https://www.omicsonline.org/open-access/threats-of-agricultural-bioterrorism-to-an-agro-dependent-economy-what-should-be-done.php?aid=27096. Cited 21 Sept 2020
19. Croddy EA (2005) Weapons of mass destruction volume I: chemical and biological weapons. Santa Barbara, CL, ABC-Clio
20. Tenenbaum DJ (2008) Food vs. fuel: diversion of crops could cause more hunger. Environ Health Perspect 116:A254 [Internet]. National Institute of Environmental Health Sciences. Available from: /pmc/articles/PMC2430252/. Cited 23 Aug 2022
21. Ryan L (2018) Agro-warfare: the basics. SOFREP [Internet]. Available from: https://sofrep.com/news/agro-warfare-basics/. Cited 23 Aug 2022
22. Ban J (2000) Agricultural biological warfare: an overview. Off Just Programs 8 [Internet]. Available from: https://www.ojp.gov/ncjrs/virtual-library/abstracts/agricultural-biological-warfare-overview. Cited 23 Aug 2022
23. Olsen A (2003) U.S. moves towards biological warfare in Colombia. PANNA [Internet]. Available from: https://www.panna.org/legacy/panups/panup_20030124.dv.html. Cited 23 Aug 2022
24. Kaur T (2019) Agro-warfare: attack on crops and livestock. CBW Mag 12 [Internet]. Available from: https://idsa.in/cbwmagazine/agro-warfare-tkaur. Cited 23 Aug 2022
25. Kageyama B (2021) Why Winston Churchill stopped operation vegetarian. Hist Yesterday [Internet]. Available from: https://historyofyesterday.com/why-winston-churchill-stopped-operation-vegetarian-e1c0a2176f04. Cited 24 Aug 2022
26. Rosie G (2001) UK planned to wipe out Germany with anthrax. Glas Her Sun, Glasgow, 14 Oct 2001
27. Tooze A (2006) The wages of destruction. Allen Lane, London
28. Graziosi A (2004) The Soviet 1931–1933 famines and the Ukrainian Holodomor: is a new interpretation possible, and what would its consequences be? Harv Ukr Stud 27:97–115 [Internet]. Available from: https://www.jstor.org/stable/41036863#metadata_info_tab_contents. Cited 23 Aug 2022
29. Phippen J (2016) Kill every buffalo you can! Every buffalo dead is an Indian gone. The Atlantic [Internet]. Available from: https://www.theatlantic.com/national/archive/2016/05/the-buffalo-killers/482349/. Cited 23 Aug 2022
30. Unknown (2022) SEAL history: underwater demolition teams in the Korean war. National Navy UDT-SEAL Museum [Internet]. Available from: https://www.navysealmuseum.org/naval-special-warfare/seal-history-underwater-demolition-teams-in-the-korean-war. Cited 23 Aug 2022
31. Rogers P, Whitby S, Dando M (1999) Biological warfare against crops. Sci Am 280:70–75
32. Keremidis H, Appel B, Menrath A, Tomuzia K, Normark M, Roffey R et al (2013) Historical perspective on agroterrorism: lessons learned from 1945 to 2012. Biosecur Bioterror 11:S17–S24 [Internet]. Mary Ann Liebert, Inc. Available from: https://www.liebertpub.com/doi/10.1089/bsp.2012.0080. Cited 30 Nov 2022
33. Kambouris M (2021) The concept of humanome and the microbiomic dimension. In: Kambouris M (ed) Genomics in biosecurity, 1st edn. Elsevier Academic Press, London, pp 15–28
34. Kambouris ME (2023) Global catastrophic biological risks in the post-COVID-19 world: time to act is now. OMICS 27:153–170 [Internet]. Available from: https://pubmed.ncbi.nlm.nih.gov/36946656/. Cited 24 July 2023
35. Owens S (2001) Salt of the earth. Genetic engineering may help to reclaim agricultural land lost due to salinisation. EMBO Rep Engl 2:877–879
36. Gribanova GI, Karatueva EN (2022) The specific definition of ecoterrorist organizations in the USA and Russia. Acad J Interdiscip Stud 11:41–55
37. Smart JK (1997) History of chemical and biological warfare: an American perspective. In: Sidell FR, Takafuji ET, Franz DR (eds) Medical aspects of chemical and biological warfare. Borden Institute, Washington, DC, pp 9–86
38. Smits DD (1994) The frontier army and the destruction of the buffalo: 1865–1883. West Hist Q 25:312

39. Gisselsson D (2022) Next-generation biowarfare: small in scale, sensational in nature? Health Secur 20:182–186
40. Flade F (2018) The June 2018 Cologne ricin plot: a new threshold in jihadi bio terror. CTC Sentin 11 [Internet]. Available from: https://ctc.westpoint.edu/june-2018-cologne-ricin-plot-new-threshold-jihadi-bio-terror/
41. Palmer M, Tiu B, Weissenbach A, Relman D (2017) On defining global catastrophic biological risks. Health Secur 15:347–348 [Internet]. Available from: https://pubmed.ncbi.nlm.nih.gov/28737976/. Cited 6 Sept 2021
42. Manheim D (2018) Questioning estimates of natural pandemic risk. Health Secur 16:381–390 [Internet]. Mary Ann Liebert Inc. Available from: https://www.liebertpub.com/doi/10.1089/hs.2018.0039. Cited 2 Mar 2021
43. Millett P, Snyder-Beattie A (2017) Human agency and global catastrophic biorisks. Health Secur 335–336
44. Schoch-Spana M, Cicero A, Adalja A, Gronvall G, Kirk Sell T, Meyer D et al (2017) Global catastrophic biological risks: toward a working definition. Health Secur 15:323–328
45. Yassif J (2017) Reducing global catastrophic biological risks. Health Secur 329–330
46. Nikolić SS (2017) An innovative response to commercial UAV menace-anti-UAV falconry. Vojn Delo 4:146–167 [Internet]. Available from: http://strangevehicles.greyfalcon.us/Nazi%20Pigeon.htm. Cited 5 Aug 2022
47. Özdemir V (2018) The dark side of the moon: the internet of things, industry 4.0, and the quantified planet. Omics J Integr Biol 22:637–641 [Internet]. Mary Ann Liebert, Inc. Available from: https://www.liebertpub.com/doi/10.1089/omi.2018.0143. Cited 10 May 2019
48. Navas-Molina JA, Hyde ER, Sanders JG, Knight R (2017) The microbiome and big data. Curr Opin Syst Biol 4:92–96 [Internet]. Elsevier. Available from: https://www.sciencedirect.com/science/article/pii/S2452310017301270. Cited 10 May 2019
49. Özdemir V, Hekim N (2018) Birth of industry 5.0: making sense of big data with artificial intelligence, "the internet of things" and next-generation technology policy. Omics J Integr Biol 22:65–76 [Internet]. Available from: http://www.liebertpub.com/doi/10.1089/omi.2017.0194. Cited 10 May 2019
50. David L, Brata AM, Mogosan C, Pop C, Czako Z, Muresan L et al (2021) Artificial intelligence and antibiotic discovery. Antibiotics 10
51. Newcomb T (2023) Global catastrophic cyber event coming in 2 years, report says. Pop Mech [Internet]. Available from: https://www.popularmechanics.com/technology/security/a42660926/global-catastrophic-mutating-event-coming-in-2-years/. Cited 7 Sept 2023
52. Manheim D, Lewis G (2021) High-risk human-caused pathogen exposure events from 1975–2016. F1000Res
53. Lentzos F (2012) Synthetic biology, security and governance. Biosocieties 7:339–351 [Internet]. Available from: http://www.ncbi.nlm.nih.gov/pubmed/32226471. Cited 21 Sept 2020
54. Gómez-Tatay L, Hernández-Andreu JM (2019) Biosafety and biosecurity in synthetic biology: a review. Crit Rev Environ Sci Technol 49:1587–1621 [Internet]. Taylor and Francis Inc. Available from: https://www.tandfonline.com/doi/full/10.1080/10643389.2019.1579628. Cited 18 Sept 2020
55. Akolpoglu MB, Alapan Y, Dogan NO, Baltaci SF, Yasa O, Tural GA et al (2022) Magnetically steerable bacterial microrobots moving in 3D biological matrices for stimuli-responsive cargo delivery. Sci Adv 8:eabo6163 [Internet]. American Association for the Advancement of Science. Available from: https://www.science.org/doi/10.1126/sciadv.abo6163. Cited 6 Oct 2022
56. Steele EJ (2020) Introduction—panspermia, 2020. Adv Genet 106:1 [Internet]. Elsevier. Available from: /pmc/articles/PMC7340401/. Cited 29 Nov 2022
57. Steele EJ, Gorczynski RM, Lindley RA, Tokoro G, Temple R, Wickramasinghe NC (2020) Origin of new emergent coronavirus and Candida fungal diseases-terrestrial or cosmic? Adv Genet 106:75–100 [Internet] Elsevier. Available from: https://pubmed.ncbi.nlm.nih.gov/33081928/. Cited 29 Nov 2022

58. Schmidt M (2010) Xenobiology: a new form of life as the ultimate biosafety tool. BioEssays 32:322–331 [Internet]. Available from: https://pubmed.ncbi.nlm.nih.gov/20217844/. Cited 21 Sept 2020
59. Green MS, LeDuc J, Cohen D, Franz DR (2019) Confronting the threat of bioterrorism: realities, challenges, and defensive strategies. Lancet Infect Dis e2–e13
60. Barrow P, Dujardin JC, Fasel N, Greenwood AD, Osterrieder K, Lomonossoff G et al (2020) Viruses of protozoan parasites and viral therapy: is the time now right? Virol J 17 [Internet]. Available from: https://pubmed.ncbi.nlm.nih.gov/32993724/. Cited 20 Aug 2022
61. Těšitel J, Li AR, Knotková K, McLellan R, Bandaranayake PCG, Watson DM (2021) The bright side of parasitic plants: what are they good for? Plant Physiol 185:1309–1324 [Internet]. Available from: https://pubmed.ncbi.nlm.nih.gov/33793868/. Cited 29 Nov 2022
62. Nazir F, Malik M, Hussain M, Asif N (2017) Insects as biological weapons. J Bioterror Biodef 09
63. Keiichi T (2005) Unit 731 and the Japanese Imperial Army's biological warfare program. Asia-Pac J 3:2194 [Internet]. Available from: https://apjjf.org/-Tsuneishi-Keiichi/2194/article.html. Cited 3 Nov 2020
64. Borges do Nascimento IJ, Pizarro AB, Almeida JM, Azzopardi-Muscat N, Gonçalves MA, Björklund M et al 2022 Infodemics and health misinformation: a systematic review of reviews. Bull World Health Organ 100:544–561

Chapter 2
The Biosecurity Field so Far

Abstract Technology has offered considerable options and an impressive and multi-level array of means to create, enhance and upkeep biosecurity during the last quarter-century. The diagnostics, such as Molecular/genomic analysis and scanning methods, with the Nucleic Acid Amplification Tests being in the forefront, or the Serodiagnostics that now come regularly in solid formats as Rapid Diagnostic Tests but also improvements in Microscopy and culturing through the advent of Culturomics are undeniably the mainstays of this progress. Similarly explosive advances refer to all the issues incorporating Bioinformatics and remote applications, such as lab and diagnostic automations, electronic health files and their fast turnover and dispatch, remote examination and test analysis. Moreover, intervention by therapeutic means seems also to be evolving despite the scarcity of new antibiotics. New micro- and macromolecular antimicrobial compounds of biological origin and whole organisms such as bacteria and viruses/phages, along with advanced and controllable delivery methods are coming online for normal medicinal use and may be adopted, adapted and modified for Biosecurity use. However, most of these modalities regard the biothreats as simply a perpetrated case of a natural epidemic or, rather outbreak, and are thus susceptible to perpetrators' suppressive or deceptive countermeasures.

Keywords UV-LIF · DNA microarrays · Monoclonal antibodies/antibody fragments · Antibiotics · LIDAR · Bioaerosol · Biopharmaceuticals · Virotherapy · Bacteriotherapy

The current iteration of Biosecurity has been—grudgingly—updated due to COVID-19, but is still lagging behind both current and projected threats. It could be seen as a construct of two major domains: Biosecurity proper on one hand, and various other developments that enable and uphold Biosecurity. The latter consist a very heterogeneous group, where one may find the cyber sector tasked to tackle cyberthreats of biosecurity applications, such as Infodemic (i.e. agentless and cyber-enhanced/leveraged bioevents) or fully cyber-enabled/perpetrated bioevents as detailed in Chaps. 5, 6 and 7.

M. E. Kambouris, *Preparing for the Meta-Biothreat: Disruptive Biosecurity*, Advanced Sciences and Technologies for Security Applications,
https://doi.org/10.1007/978-3-032-25690-4_2

Another sector is the use of intelligence communities to determine the identity of possible perpetrators, state-, non-state or individual players may these be. They are usually sorted by their enmity/malicious intent and filtered by their capacity-capability, in a manner fully compatible with security services' *modus operandi*. The sector is a bit expanded, though, similarly to the nuclear/radiological control regimes, by trying also to identify possible agents that may be used, as well as the methods of delivery, including deployment, transportation, principle of operation, technology, processes and procedures and training/recruitment of operatives. Once more the objective is similar to standard defense and security/secret services interests and operations, but in here the inbred dual-use in almost all kinds of organisms perplexes the issue, especially once Synthetic Biology achieved optimized production and modification procedures. Actually, the Security establishments seem unable, in this phase, to keep pace with the explosive development of biosciences and the criminogenic or other nefarious and malicious potential these developments carry. In fact, as will be detailed in Chaps. 3 and 4, they are *all* bound to be weaponized or otherwise mis-/abused….

The other domain is a more cohesive, although still loose amalgamation of Medicine (or rather Health Upkeep sciences), Biology and Public Health, the latter including different iterations and some parts of Ecology, such as Urban Ecology. This domain uses technology, procedures and knowledge/expertise to prevent and react to the infection event, in the environment (a prerogative of the Public Health) or on/within the targeted organisms (a prerogative of Medicine, but also of veterinary science and plant protection). It uses amenities (1) to disrupt dissemination, by destroying the agents before entry; this includes cases where they are free in the environment (soil, air, surfaces/objects, water) or on/in different vectors and hosts. It is an aggressive approach, striving to resolve the threat fully and permanently, and was pursued by the Chinese authorities during the COVID-19 [1–3]. (2) To keep the targets but also vectors physically free from the Pathogens, by restrictive measures, as the use of Personal Protective Equipment and procedures such as washing, decontamination, sterilization [4–6]. Compared to (1), this is a much more defensive approach, very taxing in the long term but less severe in implementation.

The next step (3) regards the total or partial failure on any of the former ones to prevent infection, or the anticipation of such failure. It includes all known medical interventions ("medical" taken *sensu lato*, to include plants, animals and other bioentities, such as ecosystems and biomes), usually described under the term Medical Counter- Measures (MCM) although this might be slightly restrictive [7–10]. And step (4) informs the three previous ones as it detects and identifies in spatiotemporal, biological and biochemical terms the existence and nature of the threat. It is the realm of theoretical and applied biosciences, bolstered by other natural and computational sciences, a truly multi-disciplinary complex of engineering, technology, knowledge, cognition and know-how, which form the Biosecurity concept and other similar and cousin ones, such as Bioresilience [11–13].

Regarding the step (3), in a medicinal context, before the main, reactive, therapeutic treatment, there are two proactive substeps: to prepare the target organisms by improving their health and welfare, making them more robust to resist to infection, symptoms development and rehabilitation after clearing the pathogen [14, 15], and to train their immunity with vaccination [16, 17]. In some cases a proactive, prophylactic distribution of antidotes/antimicrobials may be needed, to create usable levels in terms of (pharmaco)kinetics and especially so to individuals/populations with weak health and/or immunity [18–20].

In this step there is a complicating issue, in cognitive terms: a *proactive* medicinal approach (for example vaccination, which is implemented *before* infection and in order to forestall or alleviate it) may be a *reactive* approach in terms of Biosecurity. Vaccination needs a clearly defined threat to design the vaccines against, and a high probability of occurrence to initiate such measure, which is by definition fraught of uncertainties, dangers and high cost. Consequently, it is designed and implemented *after* a biothreat is detected, characterized and prioritized, thus consists a *reactive biosecurity* measure, but is applicable to as yet undiseased and actually uninfected subjects, thus a *proactive medical* intervention. Such complications exist regularly where interfaces of different disciplines and sub-disciplines occur and may cause havoc under pressure, obstructing the cohesive thought, discussion and collective cognitive effort to react to a crisis; so, such issues must have been identified and resolved beforehand, not in the context of *ad hoc* conglomerates of experts recruited when a crisis is imminent or actively developing.

2.1 The State-of-the-Art in Intervention

The reactive medicinal intervention is the treatment of diseased individuals or populations, which may be directed in tackling the causative agent [21], the prerogative of the "causal therapy" [22]. However, in perpetrated bioevents this approach may be tricky, especially if the agent is unknown or novel; in such cases the option is the "symptomatic" or "supportive" treatment/therapy [22–25], to stabilize the patient and alleviate the symptoms. This approach may be key to survival in general [22, 23], but also as a factor of temporal importance: support of the diseased individuals until their immunity may react effectively, or administered causal therapy becomes measurably effective, or, *in extremis*, new therapies of causative nature are developed, produced and distributed, or existing ones are tested for cross-protection and retasked [26]. Such therapies would include terminal effector moieties, such as sera, blood plasma, whole blood etc. [22–25] but also bioactive compounds, such as anti-inflammatory agents or blood pressure regulators. Obviously, massive severe events cannot be treated in the abovementioned fashion—as obvious with the COVID-19 pandemic [27], as they exceed the health provision capacities in health- support systems based on accurate cost-effect projections.

2.1.1 Antibiotics

Then come the causal therapy modalities. The mainstay are the antimicrobials, a category that encompasses substances of extremely diverse composition, nature, mode of action and intended result. The antibiotics are micromolecular drugs of natural origin but produced synthetically, modified or not. They may target one specific kingdom of pathogens: viruses (antivirals), bacteria (antibacterials), fungi (antifungals), proto/metazoan parasites (antiparasitics). The antidotes that are expected to counteract toxins/bioregulators are not among the antimicrobials, by any notion [7]. The antibiotics are characterized by their elaborate mechanism of action, and this also includes purely synthetically designed and engineered such amenities, which are usually termed "antimicrobial (bio)chemotherapeutics". If an antibacterial or antifungal is expected to stop the propagation of the targeted microorganism, it is considered a -static compound (bacteriostatic, fungistatic); if they cause necrosis of the microbial cell, they are -cides (bacteriocides, fungicides). The antiparasitic drugs usually avert the adhesion rather than actively stop propagation or cause necrosis, thus they are antiparasitics without further ado [28–30].

For the antivirals [26] the distinction of -cides and -statics is not applicable either, at least at the moment; but it should be. They are now considered lifeforms, acellular in nature [31]. Thus, not allowing the adherence of the virion on its target cell is very different from damaging the virion proper, or the contained or delivered nucleic acid. The latter means that the given virion/provirus is permanently incapacitated, while, if adhesion is thwarted, the virion may become active infectant in some other set of conditions. Such cases are if it jumps to another host or environment, or if the suppressive drug concentration/levels drop. Occasionally the "-statics" might turn "-cidal" and vice versa, depending on dosage. For example. the partial destruction/deterioration of the cell wall of a bacterium or a fungus might incapacitate it and stop its propagation, as it strives to repair the damage, or the damage may be severe enough so that the cell is necrozed [32].

Not all classes of antibiotics have such scalability, nor is it always predictable, as there are too many parameters of both dynamic and kinetic nature. It must be underlined though, that the -cidal is not always the better way to go; it might entail toxicity for the host, which becomes more severe and probable with the relation of host and pathogen. If an Eucaryote host takes antifungals, which target Eucaryota pathogens, extreme toxicity may appear if the drugs target common or cousin (closely related) mechanisms, as is the case with the Amphotericin-B antifungal agent, which acts on fungal membranes' ergosterol but also with animal cell membranes' cholesterol [33]. Additionally, a microbicidal agent entails toxicity for the hosts' appendage microbiomes [34], which might cause numerous complications including immediate or short-term ones, such as a drop in immunity or loss of fitness and general health [35]. There are always issues of antimicrobial stewardship [36], too, but in outbreaks, issues of availability might arise as well [20].

The use of antibiotics balances between narrowly targeted, very specific ones, and more widely targeted, wide-spectrum ones. The latter are far better if accurate identification is not at hand, as is in empiric therapy or in cases of emerging/novel pathogens. The narrow spectrum ones are best option for their intended targets, which is of some practicality in natural, spontaneous outbreaks, but occasionally inapplicable in perpetrated biothreats, as the perpetrator might have factored such capabilities, which are no secret of the state, in the selection of the offensive pipeline. True must be told, that in many cases, and especially in cyber-leveraged events, such narrow-spectrum formulations might be the best course of action.

The micromolecular moieties of either natural or synthetic origin, massively produced in synthetic (bio)chemistry setups was the quintessence of the Antibiotics Revolution that would have eradicated microbial diseases [37]. It fades away, as Antimicrobial Resistance (AMR) increases by the month [36], very few new antibiotics come into the pipeline [38] and this is only partially an issue of biological evolution [39]: the promise of Pharmacogenetics in the early 2000s for new drug designs [37, 40, 41] (proven void up to now) made many companies close their chemical chemistry research labs, outsource research and development to universities and actually stop operations in the field due to low profitability, compared with other sectors [42].

This shows exactly how erroneous and misguided is any strategic inclusion, from a safety/security point of view, of the private sector to vital health needs, as if market manipulations and not-best-practices, such as preferentially selling to the more promising- or threatening-customer were not enough. It Is possible if not probable that if public establishments take over the research for antibiotics the shortage will be resolved [43]; after all, alleviating measures where the public pre-pays and the private industry reaps high profits and low risk, such as financing tools like product development partnerships (PDPs) and especially Push Funding [44] and Options Market for Antibiotics (OMA) [45], show that the problem is less of Biology than usually suggested. Be it as it may, there is a clear tendency by the industry to invest in vaccines, preferably of executable nature (xNAs) for alleged developmental cost reasons [46] ("alleged" as the reason for the exorbitant price tag for developing conventional vaccines, compared to decades past with lower technology and knowledge levels, has not been addressed convincingly), but this is no biosecurity policy; it may only apply to *known* pathogens.

2.1.2 Biologicals/Biopharmaceuticals/Microbials/Probiotics

The other reactive approach that seems promising and receives attention in the struggle against infectious diseases is the combined field of amenities produced directly from bioentities. These may be molecular compounds, both of micro- and macro-molecular categories, and whole living organisms. The nomenclature remains a bit vague up to date. Still, it may be maintained that the Biopharmaceuticals or Biologicals are medicines the active drug compound of which is produced by living cells. This compound usually (but not necessarily) is a macromolecule rather than a micromolecule and thus much more sensitive in external conditions throughout its product life cycle [47–49]. On the other hand, Pharmabiotics would be the proper term for organisms used as medicament formulations, much like the probiotics, and it would be tempting to name thus probiotic formulations with directly therapeutic bioactivity [50], although the term is ambivalent and not solidly defined [51]. Whether engineered/modified such compounds produced by engineered or fully artificial cells or organisms would still qualify as Biopharmaceuticals is a question, but most probably the degree of engineering is irrelevant once the production occurs from living cells/machines, which seem to be termed Microbials [52]. In vitro production settings, such as cell-free bioreactors, on the other hand, might not qualify: they verge to fully synthetic production, albeit by different methods than the synthetic (bio)chemistry, but in a very similar context. The Biosimilars are sufficiently defined as the equivalents of generic drugs produced once the patent of a biopharmaceutical expires [47].

The monoclonal antibodies, the antibody fragments, recombinant or not [52], and other such compounds along with insulin [53] are the most widely recognizable biopharmaceuticals [47, 48], but antimicrobial peptides may become instrumental [54, 55]. The monoclonal antibodies are high-specificity antimicrobial effectors that may prove an extremely effective, quick-reaction although expensive stop-gap solution as in the COVID-19 pandemic [56], since they may be developed and produced in an agnostic setup and, consequently, most expediently. Additionally, the same moieties may be used for highly specific, targeted diagnostic tests, in both wet and dry Rapid Diagnostic Test (RDT) formats, once more as in the COVID-19 self tests [57].

The most promising biologicals, though, are the microbial therapies, both in Public Health/routine Medicine and in Biosecurity formats. There are viral concoctions of phages for virotherapies against bacteria [58] and of oncolytic viruses against tumors [59, 60]. More applications are coming, of Mycoviruses against fungi [61, 62], parasites [59] and against other viruses due to the advent of the virophages [31, 63, 64]. Fungi may be used to deter or clear bacteria and other fungi [65], as many antibiotics originate from natural fungal antibacterial compounds [66]. Bacteriotherapies may be used against fungal [67] and bacterial pathogens [68] and against tumors [69].

Although Cancer supposedly is not included in the perpetrated biothreat scope, this may be erroneous. Dirty bombs [23] and other radiotoxins such as Polonium, Thalium [70], but also residual radioactivity, tumorigenic viruses—oncoviruses [71], tumor-inducing bacteria [72, 73], and chemical agents [74] are all instruments of possible biowarfare/terrorism/crime attempts that may cause tumorigenesis and with no small measure of deniability for any perpetrator. Consequently, the ability to treat such conditions affordably, and with as little toxicity for the host as possible, is obviously appreciable, as a fail-safe in the case such agents would be deployed successfully. Moreover, it may prove to be a deterrent for the use of these particular biorisk agents, as it would negate or undermine their effectiveness.

2.2 The State-of-the-Art in Surveillance

All the above proactive or reactive intervention schemes need guidance. Intervention must be informed, else it may be counterproductive, to say the least. The example of the different vaccines against Smallpox, the scarecrow of the 90s in terms of Biosecurity, is revealing. The perceived threat opened a vast market for no actual reason, with the simple insinuation that the eradicated disease *might* have been weaponized by rogue states/agents despite the firm control over its two (2) surviving strains. The closely kept strains allowed some vaccines to be produced urgently in the West, which were understood that would have adverse effects in up to 10% of the receiver population, and thus any mass purchase of the vaccine was mandatorily followed by a complementary purchase of an antidote (the term is used *sensu lato*, as was at the time) [75]. The East, having weaponized Smallpox [75, 76], and thus having produced great volumes of relevant vaccines, somehow did not consider the adverse reactions of any importance so as to take measures. Whether this was a calculated risk, or they did not expect any reverse effects of any scale is up to conjecture. The thing is that a vaccine produced at ease with modern technology for a traditional, unmodified agent [77] that should have not existed as a threat (else the whole structure of Conventions, tests, Selection lists had been an expensive and disinformative failure) was expected to present to up to 10% severe, life-threatening adverse effects. It is a matter of conjecture if within the chaos of a perpetrated Smallpox outbreak (it could not happen spontaneously, the disease was eradicated) with possibly modified strains one would have the luxury to attend to such details and use medical resources for a low percentage of non-transmittable complications. But the fact remains that needless vaccinating for the Smallpox with the new, advanced western compounds would cause 10% of health casualties if not more. Intervention *has to* be informed.

And this is the task of Biosurveillance; to suggest and inform intervention or opt for idleness. It is multi-level and complex a task, combining and reconciling diverse needs, technologies, cognition levels and applications. It is obvious that there are at the very least three key factors: spatiotemporal (where and when), which is an external one, subjective (perpetrators, controllers, enablers) and operational: this is once more very complex an issue, as it refers to the intended biotarget (human, company animals, livestock, environment, agriculture), and to the agent/pathogen, including its deployment/dissemination mechanism and method, any possible enhancements and updates, its unitary or collective nature (pathogens-plural rather than one pathogen), but also to further details of interest. Such may be, indicatively, reservoirs, source host or vector species and populations of origin, phylogenetically related groups, populations and taxa, or bearers of specific biomarkers/characters.

2.2.1 Diagnostics

If Surveillance is defined as the observation of the known horizon of events to apperceive any anomaly, and Diagnostics is the understanding and determination, simply put, of a malfunction in a given physiology, there is a certain overlap. Some contacts/anomalies the surveillance may pick could well be the causality of derailed physiology. This is not exclusively so; some anomalies picked by surveillance may be benign, and ailments may skip surveillance, by being hereditary or idiopathic [78, 79].

There are many different levels of (Bio)Surveillance, in part due to the different professionals and scientists that are called to manage such crises and threats, thus complicating communicating among the responders of different backgrounds. The medical dimension is the Symptomatic Surveillance, where signs (objectively observed by the attending physician) and symptoms (subjectively reported by the patient) are recorded and communicated so as to determine an outbreak by spatiotemporal surge tracks. It is obvious that this approach detects and processes not the bioagents themselves, but their effect, which is enough for symptomatic response but not always so for causal therapy, as many agents might cause similar symptomatology, not to mention modified or novel ones, which cannot be identified indirectly in this manner. Additionally, the symptoms of one agent may vary in different hosts of the same species, due to variable genetic composition.

The biological dimension is to seek biomarkers by a multitude of laboratory or field tests; the latter being simplified versions of the former, most of the times. It is actually an effort to detect directly the bioagents themselves, not their result and impact, but it still applies best in unmodified agents. Modified ones may have altered biomarkers and other signatures, exactly to thwart identification, classification and well-informed, adapted response. Still there is a range of tests, laboratory and field ones, the latter prompted by the COVID-19 experience, that may identify the agents in internal and external biocompartments and fluids/secretions of the biotarget, but also in the environment, including but not limited to solid objects, environmental liquids, soil and live carriers/vectors. It is much easier to detect bioagents in absence of chemical or biological background, but this is a rare luxury, and in perpetrated events it may be a deliberate occurrence. It is a boon even to work with clearly heterologous biological background, as when detecting microbiota in sterile biocompartments, which is the essence of Clinical Microbiology, both in conventional, serological and molecular terms.

The biological iteration of biosurveillance tries to answer some questions, either collectively—"polyphasically" [80, 81], i.e. by a panel of techniques applied sequentially but occasionally complementarily; or individually—"monophasically" [82], whence one technique answers all questions of interest. Such are the question of suspect biopresence, i.e. whether there is something somewhere (detection); of its nature and identity (identification); and of any causal relation with some ailment. The latter relation may be prospective, meaning the pathogenic potential, or actual, regarding the pathogenicity of the detected- and possibly identified-contact in the context of the observed symptomatology (microbiological/laboratory diagnosis). The last step consists the interface of biology with clinical microbiology.

2.2.1.1 Conventional Detection and Identification

The term "Conventional Microbiology" covers the knowledge, cognition, methodology and techniques that were mainstream before Serology and Molecular Biology/ Nucleic Acids research in Microbiology ("Molecular Microbiology") coming on line. Although it is diverse a sector, it basically refers to cultures and microscopy (usually visual microscopy) to detect and study a microbiote [83]. These two techniques are complementary.

Microscopy offers a direct visual trace of the microbiote and its parts, organelles and perhaps macromolecules. Based on the image collection and transportation medium, there is the visual microscopy, using light [84], electronic microscopy, using electron beams [85], and the scanning probe microscopy, of nanometer-scale resolution, able to observe—or actually feel—micromolecular entities [86], while X-ray microscopy with atomic scale resolution seems to emerge as a fourth category [87]. These three major technological steps are used for different sizes of interest: respectively cells (micrometric analysis), sub-cellular/multimolecular level and organelles (sub-micrometric/nanometric analysis), single-though large-molecules, like a single protein structure (single-digit nanometric analysis) and small molecules/submolecular level. In these categories many different techniques and concepts kept emerging during the last 3 centuries. As most cells and their components are whitish and transparent, in visual light microscopy the use of dyes, these being colored compounds [88], to enhance the observable characters, is established if not imperative. Such respective amenities in electron microscopy are heavy metals such as Gold [89]. Microscopy of different basic technologies, using non-poisonous dyes, combined with video and other techniques of signal processing allows observation of the cells being live and thus sequences of snapshots revealing the cellular activity, such as motion and flow, are made redundant [90, 91].

The culture is the paramount proof that a microbe (or a cell in general) is present, alive and able to grow and—conditionally—to produce spawn. Microscopy proves its cellular existence, but not its virility, although differences in membrane permeability to some dyes show if the cell is alive or necrozed [92] and scanning probe and electron microscopy, as well as automated video-recording visual microscopy, substantiate sub-cellular motion of organelles and chemical compounds, which are a proof of survival but not of proliferation. The culture proves existence, survival, health and virility and offers increased quantity of sample for a number of subsequent analyses, either cultured-based, such as physiology and metabolic tests and resistance to antibiotics, or non-culture based, such as microscopic examination, molecular/genomic analysis, serological determination for identification or assay development and, finally, for spectroscopic analysis; not to mention direct observation of macroscopic biomarkers, such as growth patterns and colors [93]. The use of cultured samples for characterization and identification bypasses quantitative/

sensitivity issues that plague many tests, especially serological and biochemical, and also enriches and occasionally purifies the sample, a prerogative for spectrometric analysis and a highly appreciated assistance in genetic/genomic analysis as it diminishes contamination and noise [94].

Culture is time-consuming and this fact made molecular methodology, the Culture-Independent Diagnostic Tests (CIDT) of choice of the mid-90s, a harbinger of progress in Medicine. The time for identification of an agent dropped to hours rather than weeks, when starting from a clinical sample [95, 96]. Far worse is the fact that at least for the time being, there are many microbiota that cannot be cultured, the unculturable/fastidious microbiota [97]. Actually, the conditions and mixtures for their controlled propagation (the definition of Culture in Microbiology and Biosciences in general) have yet to be determined. The field of Culturomics has been developed to address this very fact [94, 97, 98], but in the meantime, many important microbiota remain unculturable and occasionally unknown if not revealed by Metagenomics analyses (see below) [99].

Cultures may sample objects and surfaces that cannot be processed by microscopy and serology; usually molecular methods would apply, but a degree of confirmation is welcome to the extremely high turnovers of metagenomic microbiomics (see below). The decision threshold is the fact that for a microbe to be considered a health hazard it is imperative to grow and propagate in the somewhat standard conditions of its host organism, and thus in cultures emulating such environment [98]. This is not always so, as the fastidious/unculturable pathogens are an unfortunate reality [100], but mostly a pathogen does propagate within its host to cause signs and symptoms.

2.2.1.2 Serological Detection and Identification

The development of serology has not been revolutionary in the last decades: it may be used in both diagnostics/surveillance and in therapeutics/intervention, but only for known, well-characterized threats, after considerable delay, analysis and expenditure; or for novel, even unknown agents that show cross-reactivity with known, established moieties, either of causal or of stochastic nature. It is not a quick-reaction modality, and it is not working in an agnostic setup, if not used in a highly parallel format, as in microarrays. Even then, it is not a true "agnostic-capable" method [101–103]; just one with an extended scope of discrimination. Still, this technology remains practical and important, if not vital. The main application is based on the use of either intact antibodies, such as monoclonal (but also polyclonal) preparations or of antibody fragments for direct, reactive intervention based on biologicals/biopharmaceuticals [52]. There are a number of effects these may attempt, from usual ones through complement activation and removal from the circulation or anchorage, to more novel approaches with conjugated effectors. Fragments which retain the paratope, such as Fab and $F(ab')_2$ [104] are less immunogenic by themselves, ill-suited for direct immunity interfacing and more suitable as "seeker parts" for some terminal effector; the latter may be any kind of toxin (biotoxin, chemotoxin or especially—radiotoxin) or biologicals or any other moiety, including photo/

thermoreactive ones, as are Gold particles [105, 106]. The specificity and precision of the paratope-epitope interaction [107] allows lower use of effectors for a given effect and much lower collateral damage than the functionally untargeted use of effectors, even if delivered with spatial accuracy by microsurgery/microinjection or micro/nanobot delivery as with biote-bots [108] or through the use of well-attuned pharmacokinetics considerations (taken *sensu lato*) [109].

A very similar concept is used in Diagnostics, although instead of an effector, the fragment, or the whole antibody [52, 110], is conjugated to some signal moiety—"reporter". Such moiety may be acting ("reporting") directly or indirectly. In the former category one may include radioactive compounds, producing a continuous, digital signal (Yes/No). More elaborate solutions are the dyes, with the differential, multi-color dyes being able to discriminate in a single reaction more than one target entities, thus allowing a multiplexing potential (understood as the detection of more than one different targets of interest in one sample simultaneously by the same method). The dyes may be in the visual spectrum, or beyond it, and may be standard, as used in Microscopy, or excitable, needing "interrogation" (i.e. illumination) by an energy/light source to report with transmission of a signal in a different (longer) wavelength, as in fluorescent dyes. Irrespective of their spectrum and their standard or excitable nature, dyes may be monochromatic or polychromatic (multi-color/differential, as mentioned above) [88].

In the latter category, the indirect reporting may be achieved by some functional molecule/entity, as an enzyme, which, when at the proper conditions, will produce or unmask the direct reporter, as in a color reaction, i.e. in ELISA assays [75]; other possibilities though abound. But the reporting may be achieved by some altered quality of the conjugated amenity, as in changing frequencies of absorbance or of vibrating/resonating when exited [106, 111, 112]. It is obvious that contrary to intervention moieties that are restricted to liquid state applications, even if on solid substrates of extremely small size, the reporter ones may function, in principle, in both liquid and solid state applications. The latter are the quintessence of self-tests and other RDTs in Resource-Limited Settings (RLS), such as remote area and field setups [13], allowing spatiotemporally dense patterns of surveillance and screening, virtually to the level of individuals retesting anywhere and at any time, in intervals as required by the incubation period of the syndrome; the definition of Point-of-Need Diagnostics [57, 113].

2.2.1.3 Molecular Detection and Identification—Metagenomics

Currently the different molecular techniques and approaches, meant to include the technology and science regarding the study and engineering of the nucleic acids (DNA-RNA) rather than anything else, are used in Diagnostics and Biotechnology, but not in intervention. Although the insertion of genes through Lateral Gene Transfer (LGT) and Vertical Gene Transfer (VGT) [114, 115] could be used to enhance antagonists, natural enemies and perhaps the immune capacity of a host, their use in a detrimental capacity, so as to be incorporated into the genome of a pathogen and undermine it by autotoxicosis, generic or selective silencing and other methods of genomic derailment has yet to occur, despite the potential offered by different tools but especially by the CRISPR technology [116]. The imaginative use of the recently discovered multiverse of small RNAs [117], which could be used as biologicals/biopharmaceuticals conditionally [118], seems not to be in line for tactical use in emergencies, at least not by now.

On the other hand, the technology of nucleic acids, especially—but not exclusively—the techniques using some sort of amplification (NAAT-Nucleic Acids Amplification Technology) is the mainstay of Diagnostics [119]. There are many different iterations, some purely clinical and extremely high-tech and elaborate, other ones being more adapted to tactical/emergency/expeditionary formats through methodological choices opting for robustness and sustainability rather than performance [120, 121]. Applications may be targeted or agnostic, digital or differential (including multiplex settings), wide- or narrow-spectrum, applied in PoN/RLS or in conventional or expeditionary laboratory settings [88]. The vast scope of available approaches and methods includes different breeds of thermocycling and readout/visualization technologies, occasionally executed on-site, as in solid-state polymerization or expendable, ready-to-use assay formats, such as microfluidics platforms, digital or not [119, 122], but also on-the-move, as with drone-cyclers [123] and possibly in automated UAVs/UGVs that collect, process, test and return the extracted data during their mission. Other devices have been developed for mobile or standalone operations, and even space travel-compatible models have emerged [124], within the scope of exobiology, both in terms of research and of protection for space microbiota insertion, as detailed in Chap. 3. Thermocycling is not the only option: isothermal polymerization, as is the Loop-Mediated Isothermal Amplification—LAMP approach, has been employed diagnostically since the turn of the millennium and is now advancing in digital and reverse-transcription, virus-rated iterations [125].

The PCR is by far the most popular and most differentiated NAAT methodology. The diverse PCR formats through the Real-Time PCR (RT-PCR) allow concomitant amplification and assessment of its results [126], removing the need for separate readout techniques, such as electrophoresis in agarose [95] or polyacrylamide gels [127]. In its multiplexing iteration, the PCR became capable of concomitantly processing thousands of Nucleic Acid sequences/biomarkers. It was the time of the "highly-parallel analysis", but the amplified products, in the order of hundreds to thousands of small DNA molecules ("amplicons") were beyond the capacity of any method of presenting and sorting its results, least of all the gel electrophoresis. The oligonucleotide microarrays, of roughly the same date (end of 1990s) offered the perfect method for analyzing and scoring for such massive outputs [128].

Still the combo of multiplex PCR and DNA microarrays never gained real foothold, despite its promising performance [129]. The Next Generation Sequencing (NGS) replaced the Sanger Sequencing and its spinoffs such as the four-color fluorescent dye base calling [130], which was an established, simple and efficient method, but could not handle full genomes of organisms. It took thirteen years of efforts of many labs and thousands of Sanger sequencers to produce a consensus human genome [131, 132]. The explosive three-pronged advance of bioinformatics, of DNA polymerase efficiency and affordability and of base-calling dyes and sensors prompted the NGS, which took the Sanger method to new levels of complexity and efficiency [133]. It allowed a new genome to be sequenced in a matter of days, but the complexity and cost made it suitable for Sequencing Centers: a moderate lab engaged in Molecular Biology could not afford or use it [130]. The next step, the 3rd Generation sequencing (with NGS being rebranded as "New Generation" or "Second generation" as now a Third Generation had arrived [134]), and meant to fix such deficiencies. It was meant to be less needy in Bioinformatics' resources and allow "re-democratization" of the analysis, by being small, simple and compact enough, not to mention affordable, so as to be used by different labs rather than in specialized and centralized dedicated facilities [135]. It was a perfect setup for expeditionary use. The new polymerization enzymes were much more accurate, making very few mistakes when reading/copying a DNA target, and much more efficient, producing extremely long sequence reads without being compromised or detached, which meant much less need for software- generated genome assembly, a step always fraught with uncertainties and biases. It was also much more efficient: it produced more copies of the target per some denominator. The latter was usually either Cost or any of the enzyme concentration metrics (molecules, mass/weight, Units) and less oftenly Time.

For medical applications the much faster, simpler and more affordable WES (Whole Exome Sequencing) applied the NGS principles to only parts of a genome, namely its minor part, where the protein-encoding sequences are included, wherever their topology may be [136]. There is some discussion in whether "Exome" means only the protein-encoding sequences, or also the mRNA, rRNA and trans-RNA genes, or whether the regulatory sequences of any gene should be included. Still the Biosecurity application bypasses all such concerns: WES is not enough, the true, whole, integral Whole Genome Sequencing (WGS) is required [137]. Not only due to the discovery of the small RNA molecules encoded by sequences earlier regarded as "junk", but because for Diagnostic purposes, where an engineered agent might have been used, it is imperative to have the whole genome. In this way agnostic analysis can be implemented [101–103], producing, conditionally of course, genomes that are unknown, or artificially modified (a clear case of genomic manipulation) or that should not have been present at a given site of the environment or a biocompartment of an organism. Such presence is an indication of contamination/infection, which, however, may have been spontaneous; no way to definitely characterize perpetration with the infectant alone. Modifying a genome, even to enhance virulence and other destructive qualities of the carrier cell, does not necessarily mean a deliberate release by its manufacturer.

Thus, in a medical/veterinary application, the NGS, applied in a metagenomic context (see below) is invaluable as it may fully characterize a live, genomed bioagent and show its genetic imprint, including lineage and virulence factors. This is so either in pure samples of the agent, or within sites of known genetic/sequence background, as in sterile biocompartments; the trick is that with the new technologies there is no need for culturing so as to acquire large quantities of the target genome [138]; this is the revolution of the Metagenomics [130, 139]. For non-sterile biocompartments the background is much harder to be subtracted ("filtered") by Bioinformatics approaches, and the most populous the biocompartment, the harder to do so, and the less reliable the results may be. With environmental samples, things become occasionally unmanageable: Environmental NGS-read microbiomes are *meta-microbiomes*. This means that they exist, as an output of environmental DNA (e-DNA) from a pooled harvest of microorganisms [140], only due to the assembly of sequences performed by algorithms, and not by analyzing a known and well-defined cell population. The assembly may be erroneous and biased, reshuffling existing sequences of different species to meta-chromosomes or meta-genes, meaning artificial, or rather virtual, and not existing, natural sequence constructs [141]; the end result is a metagenome that may relate to the actual collective genome of the live microbiome, or not. Thus NGS analyses in many appendage human and in environmental microbiomes may be null, and more specific and targeted, not agnostic, analyses may be the only viable option, though a not very obliging and enabling one.

The COVID-19 proceedings made undeniable a simple truth which had been overlooked intentionally, as it was very unpleasant: the overriding of tactical and scientific realities by a profit-driven health support system, which was bolstered, supported and promoted even in the middle of the crisis, was devastatingly misplaced. The curtail of expenses by consolidating testing instrumentation, supplies and expertise in few, large, thinly manned and expensively but not sufficiently equipped centers [142] in the interest of "just enough" to curtail costs and maximize profits was disastrous when faced with sudden surges in demand due to emergency (not to marketing) [143], and even worse so when supply chains miscarried, as they were similarly strained for maximum profitability [144]. There were no emergency stocks of critical protective, therapeutic or diagnostic supplies [145, 146] or surplus capacity but for very few countries. Surging of masses of potentially infected individuals in spatiotemporally restricted areas meant that the uninfected would be infected on the spot, thus the effort to respond actually increased the dissemination [147, 148]. The proper reaction was the multiple dispersed testing facilities utilizing RDT (Rapid Diagnostic Tests) [25], relatively inexpensive and thus not profitable for large diagnostic concerns and diagnostic services providers, though much so for the diagnostics manufacturing industry due to massive volumes purchased expediently. The apex was the serodiagnostic, antibody-based self-tests [149] but the definitive diagnosis at the microbiology level was performed by the much more reliable PCR tests [25, 149, 150] in expeditionary, Point-of-Need and actually Resource-Limited Settings formats [146, 150]. There had been expeditionary, portable, suitcase-transportable thermocycler models since early 2000s [126], intended mostly for military users operating under the slow realization that new rather than the few and well-known bioagents might be used in warfare or terrorism contexts [21, 151]. There were progressively lighter and more compact version of classic thermocyclers, to the point of whole Molecular Biology minilabs, of lower performance but highly deployable [152, 153]. These approaches, however, used the Peltier principle for accurate thermal steps in cyclic successions (thermocyclers) that requires solid metal heating bodies of considerable size and weight, and has high power needs, which precludes true miniaturization [154]. An impressive step were the Convection principle amplifiers [123], which were not thermocyclers: they create a temperature gradient at the two ends of a capillary containing the reaction mix and the ingredients move spontaneously between the two ends, and are subject to the reactions favored at the part of the differential they are at each time unit, thus continuously but asynchronously being subjected to the different stages of the amplification. It is obvious that PCR protocols requiring precision and carefully timed thermal steps, like the touchdown, the nested, the quantitative and the multiplex PCR iterations [129, 155–157], among others, cannot be implemented by Convection PCR/CPCR setups.

The advent of the Nanopore© concept introduced the complementarity-independent sequencing assays, whence a nucleic acid strand passes through a ring-shaped sensor that estimates their 3-D stereoelectric footprint and calls one of the respective (DNA or RNA) nucleotides [158–160], without the biases and errors inherent in complementarity-based, strand synthesis approaches, such as PCR, Sanger Sequencing, NGS, LAMP, Single Nucleotide Extension etc. Most new such platforms started being compatible with control applications running in smartphones or tablets by early 2020s [123, 160–162] rather than the laptop-hosted operating systems of previous models [124] and/or the integrated, embedded screens and control buttons [152].

The apex seems to be the Photothermal Conversion-Based cyclers, which combine the advantages of convective PCR and standard, Peltier-based thermocyclers with metallic thermal masses: the photothermal-conversion cyclers have low thermal mass but still achieve sensitivity and accuracy comparable to conventional PCR devices. Based on the orientation and conformation of the thermal energy transfer mechanisms used to heat the reaction tubes, the photothermal PCR can have a volumetric or surface planar heating mechanism [154].

The RLS compatibility is more important than usually understood. On one hand it enhances resilience, as an operational prerogative [25, 120]: it allows decentralized, distributed operations *ad hoc*, the only sustainable option in large-scale events, as in COVID-19 [25], with screening tests available for most of or for the entire population and repeatedly at that, in regular intervals. These would be executed at multiple locations, such as "hot spots" and "convenience spots". The former are areas of special epidemiological interest, like ingress/entry and egress/exit points or places of high attendance or of high vulnerability. The latter are multiple, dispersed locations, meant to facilitate the fast, uncomplicated public access and curtail the need to transit long distance and attend in high-density to access specialized and centralized testing hubs, thus alleviating the overloading, averting possible overwhelming and thus triaging of samples to be tested, and, just as important, the surge and crowd gathering in such facilities. This approach is well-served by state-of-the-art instrumentation and high- technology amenities, offering low footprint and high performance—the example being the microfluidics platforms [119, 122, 163]. Such tests may be agnostic, generic or tailor-made for a given challenge [150], albeit at relatively high cost. This might become quite an issue for massive usage with high rates of expenditure of unpredictable duration. But the problem is not just the monetary cost [145, 146]. It might be difficult to produce in volume in a surge manner, and thus accessibility and availability would be critical issues, as stockpiles for a given agent cannot be accumulated as the identity of said agents are not known when stockpiling decisions are made.

On the other hand, and fitting in the context just mentioned, the RLS compatibility also permits continued operations if supply chains are compromised- and actually leverages routine operations in underserviced communities. Remote and sparsely populated areas of the First world and backward areas in Third world countries may be serviced routinely by RLS-compatible formats as Standard Operating Procedures of Public Health services, and in these cases low-tech, which allows flexibility and low cost but at the expense of some performance, may be a practical solution: it bypasses both the monetary and the supply chain collapse-caused unavailability of expendables [164].

2.2.2 *Agnostic and Targeted Surveillance/Screening*

Irrespective of the exact technology/application involved, there are two different concepts regarding the detection-identification process: the easy one is the targeted approach, where the object of interest is known beforehand and specific tools are usually employed. One case are the specific assays which sense a given signature among many others, similar and dissimilar (i.e. pregnancy tests detecting a given hormone or tests for the SARS-CoV-2 serological and molecular biomarkers). Another case are concepts that may differentiate a more generic signal by a series of criteria and follow-on interrogations that validate the contact of interest, by eliminating similar ones. This is the case of observing a microscopic sample and differentiating it, by a number of tell-tale visual characters, from something else; or the restriction enzyme interrogation of positive PCR results on PCR-RFLP assays [95, 96, 165]. The targeted format answers a very narrow question: "Is the x factor present"? Although the x may be a broad category or a single individual, it still is a clearly defined entity, the x. These methods do not detect the z or the y but for the case of cross-reaction/reduced specificity, so one must know exactly what is expected, and perhaps implement panels of similar but different tests, as occurs with the cultures of biological fluids to determine (detect and identify) pathogens and then establish their resistance to antibiotics in Clinical Microbiology labs [29].

Another approach is the agnostic one; here the question is "what is present"—and this if it is assumed that indeed there *is* something, which means that detection must have been implemented somehow. This may have occurred indirectly, through some kind of intelligence (Medical Intelligence) or, more mundanely, by syndromic surveillance. If such primary information is not available, the question becomes much more difficult to answer: "Is there something present, and if yes, what is it?". A rather irrelevant example, but most informative, is once more pregnancy tests. They provide the initial cueing: some individual is pregnant or not, this is a detection. Afterwards, one may interrogate "what is the sex of the embryo?". That the embryo exists is a given by the previous test or tests, as self-testing for pregnancy, if positive, is usually followed by more elaborate testing, to confirm and, later, elaborate. Similarly with the Surveillance for bioagents; only this happens repeatedly and in a very wide context.

In the agnostic approach, generic biomarkers are used to detect presence and this is easier to do in microorganism-free samples, as the ones from sterile biocompartments. The agnostic approach does not provide readily actionable information, especially in high clutter from the background, as when interrogating in an environment with residual similar signal(s). Thus, seeking nucleic acids, which would suggest presence of an organism (some xenoorganisms are exempted [166], but this is a rather acceptable liability), would be next to useless in an environment bristling with Life, as is a soil sample. If the question changes to the much narrower "which are the microbes present in here?" the options and the conditions become more manageable. It is still agnostic, although narrower a search: one searches for *any* signal of a given category. A targeted search seeks a *given*, or a *number of given* signal(s).

2.2.3 Environmental/Spatial Surveillance

The existence of a bioagent in the environment, before it becomes hosted or contaminant, is a matter of public/ecosystemic/environmental health and much more relevant to military and security Surveillance formats, procedures and know-how. The methodology to detect such presence, either isolated in comparatively microbe-free environments, or within existing microbiomes in which they do not belong, requires a complex set of technical capabilities, methods and procedures. Among these, the approaches used in Diagnostics (conventional, serological, molecular) are applicable to some degree and as parts of a more complicated scheme, where sampling is of paramount importance as it informs the test to be implemented and defines the validity and applicability of the output. But, on top of the much more elaborate sampling, enacted by different kinds of samplers operating remotely and repetitively in defined but updatable patterns, the inclusion of completely different sensor amenities is of capital importance. In these, distance and multiple sources of noise and background interfere with the informative signal [13, 151].

For many years the field of surveillance in terms of "early warning" has been dominated by IR and UV sensors, operating separately or in combination [167]. There are good reasons for this: airborne dispersal is traditionally considered the optimal delivery method for many bioagents, especially traditional ones, in natural form (fungal spores [168]) or minimally engineered (*B. anthracis* aerosols [169]). It targets directly the respiratory tract of landborne and airborne animals, which is relatively exposed and its infections are potentially, directly or indirectly, lethal (especially compared to the digestive tract, another relatively exposed area). It must be remembered that aerosol analysis originated in the nineteenth century exactly as an effort to correlate the breathable air and its quality with pulmonary and other diseases [170].

Additionally, airborne particles target the most extended human organ, the skin, which may be resilient enough, but loosely attached infectants may progressively be carried away, transferred to susceptible biocompartments like the eyes, contaminate remote locations, be inhaled or swallowed at some occasion or even germinate and produce secondary and tertiary infection foci. It must be remembered that area surveillance instruments need to have a relatively long range, from meters for indoors setups to tens of kilometers for strategic coverage of cities, plantations, farms and other locations of interest; they also have to operate continuously, 24 h a day, 7 days a week, and reliably (the latter understood as with few breakdowns and false alarms); preferably unattended; and, very importantly, affordably, with low needs for attendance and maintenance and few and affordable expendables and spare parts [13, 19, 171].

Given the multitude of airborne particles, from dust and sand to carbon particles and other products of imperfect combustion, particle sensors have either to ingest air and trap the particles for some later test of viability (culture is the perfect such option, and embedded culture substrates in air samplers have facilitated this process [172], which remains time-consuming, however [173]) or somehow segregate particles. This is a two-stage process: first and foremost comes the separation of biological entities from non-biologicals, such as dust, sand and combustion products [151], which is usually dealt with through particle size and UV fluorescence comparisons [13, 19]. Second comes the segregation of different biological entities: live or hibernating cells (bacteria, bacterial spores, pollen particles, fungal conidia) and virions must be differentiated from dead matter, such as pieces of wood; and, finally, in areas and seasons of high pollination, pollen must be filtered out of microbiota, as it would cause false alarms, or allow the hiding of pathogenic airborne particles within a pollen-saturated airflow.

IR remote sensing allows detection of a cloud of an aerosol and its characterization, while UV extrapolates occasionally its particle size and vector of movement and more typically the biological or inorganic nature of the particles. Thus, a bioaerosol of possibly dangerous biologicals must be detected and located as soon as possible; in standoff ranges, if in open space (many to tens of kilometers away) by IR spectroscopy [167, 174–176]. Then, it must be characterized as such, to allow time for raising the alarm and taking protective measures first and foremost, while more detailed analysis would identify the bioparticles [19]. Living organisms have a more intense UV signature than dust and sand particles, due to some residues of their macromolecules, such as some aminoacids (e.g., tryptophan), NAD(P)H coenzymes, riboflavin and nucleic acids. The UV signal is much more intense if such molecules are excited, preferably by a focused and well-attuned laser beam which makes them transiently fluorescent; this technology is called ultraviolet laser/light-induced fluorescence (UV-LIF) [173, 177] and it is similar to the reflection of light by some surfaces, although the actual physics differ. The protein- and nucleic acid-related fluorescence in biological particles is maximized at excitation wavelengths of 266–280 nm.

The spectral images are a reliable marker of biological entities, though not of their exact nature (benign or malign/virulent) and viability (dormant, necrotic or vegetative cells) or even crude identity. Signs of active metabolism provide an excitation optimum at 340–360 nm, a sign of active metabolism and thus implying vegetative, live condition, as numerous reports suggest this wavelength region provides fluorescence information closely associated with biological molecules linked to critical metabolic activity [19]. There is also a size parameter: dust, sand and many other particles differ in size from standard airborne biological particles (though not from all, especially from ones engineered to differ) and light scattering characteristics may resolve in terms of physics, without any biological input, such issues [19]. Obviously, Advanced threats [77], fully engineered, may not abide with standard size projections informed from existing microorganisms and applicable to Traditional and Emerging pathogens, and, possibly, to most of Enhanced ones [77], as the latter are based mostly on the first two categories and are engineered further. Whether engineering may also alter the light scattering characteristics to deny classification by this method remains to be seen.

If laser emissions from a single device may be attuned to different wavelengths fast, the UV-LIF may resolve size issues, biological content and also metabolic activity [19, 151, 173] with one, polyvalent remote sensor, especially if the fluorescence returned by the bioaerosol is processed for its precise signal characteristics by advanced sensor algorithms, thus infringing in laser-induced UV spectroscopy [13]. The Light Detection And Ranging/LIDAR principle may also be implementable by polyvalent lasers and thus compute range and vector with great accuracy [177]. In this case, special testing for metabolic activity, involving ATP bioluminescence, an accurate, possibly quantitative but laborious marker of metabolic activity, which demands wet chemistry on collected samples [178] may be redundant. Similarly so, the use of IR Spectroscopy sensors (of long- or mid-IR wavelengths, excitable or not), which provide longer range cloud/aerosol detection and shape/size data [174–176] may become redundant also, although their range capabilities may not. Physics are clear: the range in IR wavelengths is always expected to outdo anything UV wavelengths may strive to attain by technological breakthroughs.

2.3 Projected Challenges

The current version of Biosecurity for both spontaneous and perpetrated biothreats is well-attuned to the Traditional and the Emerging bioagents, as described in the simplistic four-part scheme of Traditional, (spontaneously) Emerging, Enhanced and Advanced bioagents [77]. The fact that the Enhanced bioagents are based on natural (Traditional) ones suggests that this category may be included, conditionally and partially, as well. As a result, the main challenge for the said, current version of Biosecurity would be expected to be the Advanced agents, but this is not so: surprisingly, the challenge is two-fold: both the Traditional and the Advanced biothreats, as the cyber-enacted or leveraged biothreats are not under separate consideration, but only as a function of the implicated true bioagents and to the degree they are (mis)used.

The case for the Advanced agents is obvious; with many health problems, not least amongst them the Traditional and Emerging infectious diseases (spontaneous and possibly perpetrated events), priority has not been assigned to a protean field, where not even the basic nature of the threat may be known with any certainty; not to speak of identity marks and functional/pathogenic characters. Thus, since the Advanced threat is considered of low probability (see Chaps. 5 and 6) and actually unfathomable (see Chaps. 3 and 4), the effort to prepare for it is minimal, as more definite and well-defined, spontaneous threats and tangible perpetrated ones have to be dealt with first. This is, honestly, a very logical assumption and decision, and a sustainable one: such preparations are useful for the routine, everyday operations, to some degree or other.

The inclusion of the Traditional threats to the short list of the most challenging pathogen categories seems controversial as per the above. Public and Environmental Health efforts are primarily attuned to deal with such agents. However, in a biosecurity context, where the perpetrated scenario is seriously taken into account, the nature of the agent is but one threat dimension. A very important and weighting one, possibly the *most important* one, but not the *only* one. The accessibility of the agents to different and diverse groups of perpetrators is focal: thus, simple, ubiquitous pathogens, manageable by the unimpressive skillset and material and cognitive toolkit required for implementing Kitchen Microbiology, become a real headache, especially for proactive tracking and suppression: the preparation of the agents, including their acquisition and storage, requires no equipment and supplies of any rarity, complexity and peculiarity in specifications and origin/manufacture and thus prone to tracking. Even worse is the fact that this observation applies to their deployment. It is easy, and highly affordable. Both these issues create a dispersed, almost omnipresent pattern of likely foci of such engineered outbreaks [179], as will be discussed in Chap. 5.

References

1. Han Y, Xu T (2024) Chinese model in global COVID-19 prevention and control: successful experience, institutional advantages and optimal path. Heliyon 10:e25701 [Internet]. Available from: https://www.sciencedirect.com/science/article/pii/S2405844024017328
2. Mahoney JG (2023) China's rise as an advanced technological society and the rise of digital orientalism. J Chin Polit Sci 28:1–24 [Internet]. Springer Science and Business Media B.V. Available from: https://doi.org/10.1007/s11366-022-09817-z. Cited 28 July 2023
3. Wu Y, Cao Z, Yang J, Bi X, Xiong W, Feng X et al (2024) Innovative public strategies in response to COVID-19: a review of practices from China. Health Care Sci Engl 3:383–408
4. Horton L, Torres AE, Narla S, Lyons AB, Kohli I, Gelfand JM et al (2020) Spectrum of virucidal activity from ultraviolet to infrared radiation. Photochem Photobiol Sci 1262–1270. Royal Society of Chemistry
5. Fins J (2020) Pandemics, protocols, and the plague of Athens: insights from Thucydides. Hastings Cent Rep 50:50–53 [Internet]. Available from: https://pubmed.ncbi.nlm.nih.gov/32596885/. Cited 6 Sept 2021
6. Grignani E, Mansi A, Cabella R, Castellano P, Tirabasso A, Sisto R et al (2021) Safe and effective use of ozone as air and surface disinfectant in the conjuncture of COVID-19. Gases 19–32

7. Kassa J (2002) Review of oximes in the antidotal treatment of poisoning by organophosphorus nerve agents. J Toxicol Clin Toxicol 40:803–816
8. Homeland Security Presidential Directive (2007) HSPD-18-medical countermeasures against weapons of mass destruction
9. Beckham TR, Brake DA, Fine JB (2018) Strengthening one health through investments in agricultural preparedness. Health Secur 16:92–107 [Internet]. Mary Ann Liebert Inc. Available from: https://pubmed.ncbi.nlm.nih.gov/29584460/. Cited 23 Oct 2020
10. Aruru M, Truong HA, Clark S (2021) Pharmacy Emergency Preparedness and Response (PEPR): a proposed framework for expanding pharmacy professionals' roles and contributions to emergency preparedness and response during the COVID-19 pandemic and beyond. Res Soc Adm Pharm 17:1967–1977. Elsevier Inc.
11. Kambouris ME, Manoussopoulos Y, Kantzanou M, Velegraki A, Gaitanis G, Arabatzis M et al (2018) Rebooting bioresilience: a multi-OMICS approach to tackle global catastrophic biological risks and next-generation biothreats. Omics J Integr Biol 22:35–51
12. Kambouris ME, Kantzanou M, Papathanasiou B, Kostis GD, Manoussopoulos Y, Drossos P et al (2016) Towards total biothreat preparedness: expanded surveillance, joint monitoring, pooled resources and the genomic option. IJISET 3:384–389
13. Kambouris ME (2018) Mobile stand-off and stand-in surveillance against biowarfare and bioterrorism agents. In: Karampelas P, Bourlai T (eds) Adv Sci Technol Secur Appl. Springer, pp 241–255
14. Amara AA, Shibl A (2015) Role of probiotics in health improvement, infection control and disease treatment and management. Saudi Pharm J 23:107–114
15. Kourilsky P (2016) The natural defense system and the normative self model [version 1; referees: 2 approved]. F1000Research 5 [Internet]. Faculty of 1000 Ltd. Available from: https://pubmed.ncbi.nlm.nih.gov/27303629/. Cited 15 Aug 2022
16. Prausnitz MR, Mikszta JA, Cormier M, Andrianov AK (2009) Microneedle-based vaccines. In: Compans R, Orenstein W (eds) Vaccines pandemic influ. Springer-Verlag, Berlin, pp 369–393 [Internet]. Available from: https://pubmed.ncbi.nlm.nih.gov/19768415/. Cited 10 Dec 2020
17. Davoodi-Semiromi A, Samson N, Daniell H (2009) The green vaccine: a global strategy to combat infectious and autoimmune diseases. Hum Vac 5:488–493
18. Segundo GRS, Condino-Neto A (2021) Treatment of patients with immunodeficiency: medication, gene therapy, and transplantation. J Pediatr 97:S17–S23. Elsevier
19. Ho J (2002) Future of biological aerosol detection. Anal Chim Acta 457:125–148 [Internet]. Available from: https://www.sciencedirect.com/science/article/pii/S0003267001015926
20. Kunkel AA, McHugh KJ (2023) Injectable controlled-release systems for the prevention and treatment of infectious diseases. J Biomed Mater Res A
21. Cieslak TJ, Kortepeter MG, Wojtyk RJ, Jansen HJ, Reyes RA, Smith JO (2018) Beyond the dirty dozen: a proposed methodology for assessing future bioweapon threats. Mil Med 183:e59–e65 [Internet]. Available from: https://pubmed.ncbi.nlm.nih.gov/29401327/. Cited 22 Apr 2022
22. Kampmeier T-G, Hessler M, Arnemann PH, Westphal M, Seidel LM, Morelli A et al (2018) Sole causal therapy worsens outcome as compared to no therapy and combined causal and goal-directed supportive therapy in ovine septic shock. Ann Transl Med China 6:400
23. Croddy EA (2005) Weapons of mass destruction volume I: chemical and biological weapons. ABC-Clio, Santa Barbara, CL
24. Green MS, LeDuc J, Cohen D, Franz DR (2019) Confronting the threat of bioterrorism: realities, challenges, and defensive strategies. Lancet Infect Dis e2–e13. Lancet Publishing Group
25. Benda A, Zerajic L, Ankita A, Cleary E, Park Y, Pandey S (2021) COVID-19 testing and diagnostics: a review of commercialized technologies for cost, convenience and quality of tests. Sensors (Basel) 21 [Internet]. Available from: https://pubmed.ncbi.nlm.nih.gov/34640901/. Cited 24 Oct 2022

26. Khamees A, Awadi S, Al-Shami K, Alkhoun HA, Al-Eitan SF, Alsheikh AM et al (2023) Human monkeypox virus in the shadow of the COVID-19 pandemic. J Infect Public Health 16:1149–1157
27. Assefa Y, Gilks CF, Reid S, van de Pas R, Gete DG, Van Damme W (2022) Analysis of the COVID-19 pandemic: lessons towards a more effective response to public health emergencies. Glob Health 18:1–13 [Internet]. BioMed Central Ltd. Available from: https://doi.org/10.1186/s12992-022-00805-9. Cited 7 Sept 2023
28. Nankervis H, Thomas K, Delamere F (eds) (2016) Antimicrobials including antibiotics, antiseptics and antifungal agents—scoping systematic review of treatments for eczema—NCBI bookshelf. NIHR Journals Library, Southampton [Internet]. Available from: https://www.ncbi.nlm.nih.gov/books/NBK363143/. Cited 22 Apr 2022
29. Mercer DK, Torres MDT, Duay SS, Lovie E, Simpson L, von Köckritz-Blickwede M et al (2020) Antimicrobial susceptibility testing of antimicrobial peptides to better predict efficacy. Front Cell Infect Microbiol 326. Frontiers Media S.A.
30. Aung AK, Haas DW, Hulgan T, Phillips EJ (2014) Pharmacogenomics of antimicrobial agents. Pharmacogenomics 1903–1930 [Internet]. Future Medicine Ltd. Available from: https://pubmed.ncbi.nlm.nih.gov/25495412/. Cited 14 Dec 2020
31. Pearson H (2008) "Virophage" suggests viruses are alive. Nature 454:677 [Internet]. Nature Publishing Group. Available from: https://pubmed.ncbi.nlm.nih.gov/18685665/. Cited 22 Feb 2021
32. Perlin DS (2011) Current perspectives on echinocandin class drugs. Future Microbiol 6:441–457 [Internet]. Available from: https://doi.org/10.2217/fmb.11.19. Cited 11 Feb 2019
33. Kamiński DM (2014) Recent progress in the study of the interactions of amphotericin B with cholesterol and ergosterol in lipid environments. Eur Biophys J 43:453–467 [Internet]. Available from: https://doi.org/10.1007/s00249-014-0983-8. Cited 11 Feb 2019
34. Bäumler AJ, Sperandio V (2016) Interactions between the microbiota and pathogenic bacteria in the gut. Nature 535:85–93 [Internet]. NIH Public Access. Available from: http://www.ncbi.nlm.nih.gov/pubmed/27383983. Cited 17 July 2019
35. Bull MJ, Plummer NT (2014) Part 1: the human gut microbiome in health and disease. Integr Med (Encinitas) 13:17–22
36. Doron S, Davidson LE (2011) Antimicrobial stewardship. Mayo Clin Proc 1113–1123 [Internet]. Elsevier Ltd. Available from: https://pubmed.ncbi.nlm.nih.gov/22033257/. Cited 30 Sept 2020
37. Ganguly NK, Saha GK (2013) Pharmacogenomics and personalized medicine for infectious diseases. Omics Pers Med 603–635 [Internet]. Springer India. Available from: https://doi.org/10.1007/978-81-322-1184-6_27. Cited 14 Dec 2020
38. Iskandar K, Murugaiyan J, Halat DH, El Hage S, Chibabhai V, Adukkadukkam S et al (2022) Antibiotic discovery and resistance: the chase and the race. Antibiotics 11:182 [Internet]. Multidisciplinary Digital Publishing Institute. Available from: https://www.mdpi.com/2079-6382/11/2/182/htm. Cited 22 Apr 2022
39. Liu G, Catacutan DB, Rathod K, Swanson K, Jin W, Mohammed JC et al (2023) Deep learning-guided discovery of an antibiotic targeting *Acinetobacter baumannii*. Nat Chem Biol [Internet]. Available from: https://doi.org/10.1038/s41589-023-01349-8
40. Nebert DW (1999) Pharmacogenetics and pharmacogenomics: why is this relevant to the clinical geneticist? Clin Genet 56:247–258 [Internet]. Wiley. Available from: https://doi.org/10.1034/j.1399-0004.1999.560401.x
41. Odds FC (2005) Genomics, molecular targets and the discovery of antifungal drugs. Rev Iberoam Micol 22:229–237
42. Power E (2006) Impact of antibiotic restrictions: the pharmaceutical perspective. Clin Microbiol Infect 12:25–34 [Internet]. Blackwell Publishing Ltd. Available from: https://pubmed.ncbi.nlm.nih.gov/16827822/. Cited 2 Mar 2021
43. Singer AC, Kirchhelle C, Roberts AP (2020) (Inter)nationalising the antibiotic research and development pipeline. Lancet Infect Dis 20:e54–e62

44. Wasan H, Singh D, Reeta KH, Gupta YK (2023) Landscape of push funding in antibiotic research: current status and way forward. Biology (Basel) 12
45. Brogan DM, Mossialos E (2013) Incentives for new antibiotics: the Options Market for Antibiotics (OMA) model. Glob Health 9:58 [Internet]. Available from: https://doi.org/10.1186/1744-8603-9-58
46. Oğuz F, Atmaca H (2022) MRNA as a therapeutics: understanding mRNA vaccines. Adv Pharm Bull 12:274–282
47. Declerck PJ (2012) Biologicals and biosimilars: a review of the science and its implications. Gener Biosim Initiat J 1:13–16. Pro Pharma Communications International
48. Kooijman M, van Meer PJK, Moors EHM, Schellekens H (2012) Thirty years of preclinical safety evaluation of biopharmaceuticals: did scientific progress lead to appropriate regulatory guidance? Expert Opin Drug Saf 11:797–801 [Internet]. Available from: http://www.ncbi.nlm.nih.gov/pubmed/22861668. Cited 10 May 2019
49. Mantle JL, Rammohan J, Romantseva EF, Welch JT, Kauffman LR, McCarthy J et al (2019) Cyberbiosecurity for biopharmaceutical products. Front Bioeng Biotechnol 7:116. Frontiers Media SA
50. Lee E-S, Song E-J, Nam Y-D, Lee S-Y (2018) Probiotics in human health and disease: from nutribiotics to pharmabiotics. J Microbiol 56:773–782 [Internet]. Available from: http://www.ncbi.nlm.nih.gov/pubmed/30353462. Cited 10 May 2019
51. Sadeghi A, Ebrahimi M, Kharazmi MS, Jafari SM (2023) Effects of microbial-derived biotics (meta/pharma/post-biotics) on the modulation of gut microbiome and metabolome; general aspects and emerging trends. Food Chem 411:135478 [Internet]. Available from: https://www.sciencedirect.com/science/article/pii/S0308814623000948
52. Spadiut O, Capone S, Krainer F, Glieder A, Herwig C (2014) Microbials for the production of monoclonal antibodies and antibody fragments. Trends Biotechnol 32:54 [Internet]. Elsevier. Available from: /pmc/articles/PMC3906537/. Cited 30 Nov 2022
53. Baeshen NA, Baeshen MN, Sheikh A, Bora RS, Ahmed MMM, Ramadan HAI et al (2014) Cell factories for insulin production. Microb Cell Fact 13 [Internet]. BioMed Central. Available from: /pmc/articles/PMC4203937/. Cited 30 Nov 2022
54. Gordon YJ, Romanowski EG, McDermott AM (2005) A review of antimicrobial peptides and their therapeutic potential as anti-infective drugs. Curr Eye Res 30:505–515
55. Campbell Y, Fantacone ML, Gombart AF (2012) Regulation of antimicrobial peptide gene expression by nutrients and by-products of microbial metabolism. Eur J Nutr 899–907. Springer
56. Chinta S, Rodriguez-Guerra M, Shaban M, Pandey N, Jaquez-Duran M, Vittorio TJ (2023) COVID-19 therapy and vaccination: a clinical narrative review. Drugs Context 12
57. Grant BD, Anderson CE, Williford JR, Alonzo LF, Glukhova VA, Boyle DS et al (2020) SARS-CoV-2 coronavirus nucleocapsid antigen-detecting half-strip lateral flow assay toward the development of point of care tests using commercially available reagents. Anal Chem 92:11305–11309 [Internet]. American Chemical Society. Available from: https://pubmed.ncbi.nlm.nih.gov/32605363/. Cited 3 Nov 2020
58. Lin DM, Koskella B, Lin HC (2017) Phage therapy: an alternative to antibiotics in the age of multi-drug resistance. World J Gastrointest Pharmacol Ther 8:162 [Internet]. Baishideng Publishing Group Inc. Available from: https://pubmed.ncbi.nlm.nih.gov/28828194/. Cited 2 Mar 2021
59. Barrow P, Dujardin JC, Fasel N, Greenwood AD, Osterrieder K, Lomonossoff G et al (2020) Viruses of protozoan parasites and viral therapy: is the time now right? Virol J 17 [Internet]. Available from: https://pubmed.ncbi.nlm.nih.gov/32993724/. Cited 20 Aug 2022
60. Kaufman HL, Kohlhapp FJ, Zloza A (2015) Oncolytic viruses: a new class of immunotherapy drugs. Nat Rev Drug Discov 14:642 [Internet]. Nature Publishing Group. Available from: /pmc/articles/PMC7097180/. Cited 20 Aug 2022
61. Ghabrial SA, Castón JR, Jiang D, Nibert ML, Suzuki N (2015) 50-plus years of fungal viruses. Virology 479–480:356–368 [Internet]. Elsevier. Available from: https://doi.org/10.1016/j.virol.2015.02.034

62. Sutela S, Poimala A, Vainio EJ (2019) Viruses of fungi and oomycetes in the soil environment. FEMS Microbiol Ecol 95 [Internet]. Available from: https://pubmed.ncbi.nlm.nih.gov/31365065/. Cited 20 Aug 2022
63. Dutta D, Ravichandiran V, Sukla S (2021) Virophages: association with human diseases and their predicted role as virus killers. Pathog Dis 79 [Internet]. Available from: https://pubmed.ncbi.nlm.nih.gov/34601577/. Cited 20 Aug 2022
64. Desnues C, Boyer M, Raoult D (2012) Sputnik, a virophage infecting the viral domain of life. Adv Virus Res 63–89. Academic Press Inc.
65. Mishra VK, Passari AK, Chandra P, Leo VV, Kumar B, Uthandi S et al (2017) Determination and production of antimicrobial compounds by *Aspergillus clavatonanicus* strain MJ31, an endophytic fungus from *Mirabilis jalapa* L. using UPLC-ESI-MS/MS and TD-GC-MS analysis. PLoS One 12:e0186234 [Internet]. Public Library of Science. Available from: http://www.ncbi.nlm.nih.gov/pubmed/29049321. Cited 17 July 2019
66. Cueto M, Jensen PR, Kauffman C, Fenical W, Lobkovsky E, Clardy J (2001) Pestalone, a new antibiotic produced by a marine fungus in response to bacterial challenge. J Nat Prod 64:1444–1446
67. Kerr J (1994) Inhibition of fungal growth by *Pseudomonas aeruginosa* and *Pseudomonas cepacia* isolated from patients with cystic fibrosis. J Infect 28:305–310 [Internet]. Available from: http://www.ncbi.nlm.nih.gov/pubmed/7522262. Cited 11 Feb 2019
68. Serrano-Villar S, Ferrer M, Gosalbes MJ, Moreno S (2017) How can the gut microbiota affect immune recovery in HIV-infected individuals? Future Microbiol 12:195–199 [Internet]. Available from: https://doi.org/10.2217/fmb-2016-0226
69. Duong MT-Q, Qin Y, You S-H, Min J-J (2019) Bacteria-cancer interactions: bacteria-based cancer therapy. Exp Mol Med 51:1–15 [Internet]. Available from: https://doi.org/10.1038/s12276-019-0297-0
70. Nathwani AC, Down JF, Goldstone J, Yassin J, Dargan PI, Virchis A et al (2016) Polonium-210 poisoning: a first-hand account. Lancet (London, England) 388:1075–1080
71. Gaglia MM, Munger K (2018) More than just oncogenes: mechanisms of tumorigenesis by human viruses. Curr Opin Virol 32:48–59
72. Yusuf K, Sampath V, Umar S (2023) Bacterial infections and cancer: exploring this association and its implications for cancer patients. Int J Mol Sci 24
73. Cummins J, Tangney M (2013) Bacteria and tumours: causative agents or opportunistic inhabitants? Infect Agent Cancer 8:11 [Internet]. Available from: https://doi.org/10.1186/1750-9378-8-11
74. Wogan GN, Hecht SS, Felton JS, Conney AH, Loeb LA (2004) Environmental and chemical carcinogenesis. Semin Cancer Biol 14:473–486
75. Broussard LA (2001) Biological agents: weapons of warfare and bioterrorism. Mol Diagn 6:323–333 [Internet]. Churchill Livingstone Inc. Available from: https://pubmed.ncbi.nlm.nih.gov/11774197/. Cited 30 Sept 2020
76. Zilinskas R (2016) The Soviet biological weapons program and its legacy in today's Russia. Report No.: 11. Washington, DC
77. Colf LA (2016) Preparing for nontraditional biothreats. Health Secur 14:7–12 [Internet]. Available from: https://pubmed.ncbi.nlm.nih.gov/26863312/. Cited 22 Apr 2022
78. Shapira Y, Katz BSP, Gilburd B, Barzilai O, Ram M, Blank M et al (2012) Geographical differences in autoantibodies and anti-infectious agents antibodies among healthy adults. Clin Rev Allergy Immunol 154–163 [Internet]. Available from: https://pubmed.ncbi.nlm.nih.gov/21229335/. Cited 2 Oct 2020
79. Tirlapur SA, Priest L, Daniels JP, Khan KS (2013) How do we define the term idiopathic? Curr Opin Obstet Gynecol 25:468–473
80. Prakash O, Verma M, Sharma P, Kumar M, Kumari K, Singh A et al (2007) Polyphasic approach of bacterial classification—an overview of recent advances. Indian J Microbiol 47:98–108 [Internet]. Springer. Available from: http://www.ncbi.nlm.nih.gov/pubmed/23100651. Cited 13 June 2019

81. Arabatzis M, Kambouris M, Kyprianou M, Chrysaki A, Foustoukou M, Kanellopoulou M et al (2011) Polyphasic identification and susceptibility to seven antifungals of 102 *Aspergillus* isolates recovered from immunocompromised hosts in Greece. Antimicrob Agents Chemother 55:3025–3030 [Internet]. Available from: https://doi.org/10.1128/AAC.01491-10. Cited 13 June 2019
82. Irinyi L, Serena C, Garcia-Hermoso D, Arabatzis M, Desnos-Ollivier M, Vu D et al (2015) International Society of Human and Animal Mycology (ISHAM)-ITS reference DNA barcoding database—the quality controlled standard tool for routine identification of human and animal pathogenic fungi. Med Mycol 53:313–337 [Internet]. Available from: http://academic.oup.com/mmy/article/53/4/313/977005/International-Society-of-Human-and-Animal-Mycology. Cited 11 Feb 2019
83. Vyzantiadis T-AA, Johnson EM, Kibbler CC (2012) From the patient to the clinical mycology laboratory: how can we optimise microscopy and culture methods for mould identification? J Clin Pathol 65:475–483 [Internet]. Available from: https://doi.org/10.1136/jclinpath-2011-200500. Cited 11 Feb 2019
84. Popescu L, Cao Z-P (2018) From microscopy to genomic approach in soil biodiversity assessment. Curr Issues Mol Biol 27:195–198 [Internet]. Available from: http://www.ncbi.nlm.nih.gov/pubmed/28885183. Cited 3 June 2019
85. Skiniotis G (2017) A snapshot of cryo-EM. Protein Sci 26:5–7
86. Cuenot S, Bouchara J-P (2018) Atomic force microscopy: a promising tool for deciphering the pathogenic mechanisms of fungi in cystic fibrosis. Mycopathologia 183:291–310 [Internet]. Available from: http://www.ncbi.nlm.nih.gov/pubmed/29128932. Cited 3 June 2019
87. Volland J-M, Gonzalez-Rizzo S, Gros O, Tyml T, Ivanova N, Schulz F et al (2022) A centimeter-long bacterium with DNA compartmentalized in membrane-bound organelles. bioRxiv [Internet]. Cold Spring Harbor Laboratory. Available from: https://doi.org/10.1101/2022.02.16.480423v1. Cited 22 Apr 2022
88. Velegraki A, Zerva L (2020) Identifying microbiota: genomic, mass-spectrometric, and serodiagnostic approaches. In: Kambouris M, Velegraki A (eds) Microbiomics: dimensions, applications, and translational implications of human and environmental microbiome research, 1st edn. Elsevier Academic Press, London, pp 77–94
89. Osumi M (2012) Visualization of yeast cells by electron microscopy. J Electron Microsc 61:343–365 [Internet]. Available from: https://pubmed.ncbi.nlm.nih.gov/23231852/. Cited 22 Feb 2021
90. Horie K, Toda K, Nakamura T, Ideguchi T (2025) Bidirectional quantitative scattering microscopy. Nat Commun 16:9712
91. Sánchez-Orellana G, Casas-Flores S, Gutiérrez-Medina B (2019) Automated, continuous video microscopy tracking of hyphal growth. Fungal Genet Biol 123:25–32 [Internet]. Available from: https://pubmed.ncbi.nlm.nih.gov/30508595/
92. Truernit E, Haseloff J (2008) A simple way to identify non-viable cells within living plant tissue using confocal microscopy. Plant Methods 4:15 [Internet]. Available from: https://doi.org/10.1186/1746-4811-4-15
93. Kambouris ME, Goudoudaki S, Kritikou S, Milioni A, Karamperis K, Giavasis I et al (2022) Beyond the microbiome: germ-ganism? An integrative idea for microbial existence, organization, growth, pathogenicity, and therapeutics. OMICS 26:204–217 [Internet]. Available from: https://pubmed.ncbi.nlm.nih.gov/35255221/. Cited 27 Apr 2022
94. Kambouris ME (2020) Culturomics: the alternative from the past. Microbiomics 155–173. Elsevier
95. Velegraki A, Kambouris M, Skiniotis G, Savala M, Mitroussia-Ziouva A, Legakis N (1999) Identification of medically significant fungal genera by polymerase chain reaction followed by restriction enzyme analysis. FEMS Immunol Med Microbiol 23:303–312 [Internet]. Available from: http://www.ncbi.nlm.nih.gov/pubmed/10225290. Cited 13 June 2019
96. Velegraki A, Kambouris M, Kostourou A, Chalevelakis G, Legakis N (1999) Rapid extraction of fungal DNA from clinical samples for PCR amplification. Med Mycol 37:69–73

97. Lagier JC, Hugon P, Khelaifia S, Fournier PE, La Scola B, Raoult D (2015) The rebirth of culture in microbiology through the example of culturomics to study human gut microbiota. Clin Microbiol Rev 237–264 [Internet]. American Society for Microbiology. Available from: /pmc/articles/PMC4284300/?report=abstract. Cited 21 Sept 2020
98. Kambouris ME, Pavlidis C, Skoufas E, Arabatzis M, Kantzanou M, Velegraki A et al (2018) Culturomics: a new kid on the block of OMICS to enable personalized medicine. Omics J Integr Biol 22:108–118
99. Flandroy L, Poutahidis T, Berg G, Clarke G, Dao MC, Decaestecker E et al (2018) The impact of human activities and lifestyles on the interlinked microbiota and health of humans and of ecosystems. Sci Total Environ 627:1018–1038 [Internet]. Elsevier B.V. Available from: https://pubmed.ncbi.nlm.nih.gov/29426121/. Cited 21 Sept 2020
100. Baldi P, La Porta N (2017) *Xylella fastidiosa*: host range and advance in molecular identification techniques. Front Plant Sci 8 [Internet]. Available from: https://pubmed.ncbi.nlm.nih.gov/28642764/. Cited 24 Oct 2022
101. Eisenkraft A (2024) Threat agnostic approach to epidemic management using continuous remote patient monitoring. Health Secur 22:156–158 [Internet]. Mary Ann Liebert, Inc., Publishers. Available from: https://doi.org/10.1089/hs.2023.0061
102. Leiser OP, Hobbs EC, Sims AC, Korch GW, Taylor KL (2021) Beyond the list: bioagent-agnostic signatures could enable a more flexible and resilient biodefense posture than an approach based on priority agent lists alone. Pathogens 10:1497 [Internet]. Multidisciplinary Digital Publishing Institute. Available from: https://www.mdpi.com/2076-0817/10/11/1497/htm. Cited 6 Oct 2022
103. Kambouris ME, Manoussopoulos Y, Patrinos GP, Velegraki A (2019) Microbial genomics in public health: a translational risk-response aspect. In: Patrinos G (ed) Applied genomics and public health. Elsevier Academic Press, pp 131–148
104. Deyev S, Lebedenko E (2009) Modern technologies for creating synthetic antibodies for clinical application. Acta Nat 1:32–50
105. Hirsch LR, Stafford RJ, Bankson JA, Sershen SR, Rivera B, Price RE et al (2003) Nanoshell-mediated near-infrared thermal therapy of tumors under magnetic resonance guidance. Proc Natl Acad Sci U S A 100:13549–13554 [Internet]. Available from: https://pubmed.ncbi.nlm.nih.gov/14597719/. Cited 5 Dec 2023
106. Huang X, El-Sayed MA (2010) Gold nanoparticles: optical properties and implementations in cancer diagnosis and photothermal therapy. J Adv Res 1:13–28
107. Pomés A, Mueller GA, Chruszcz M (2020) Structural aspects of the allergen-antibody interaction. Front Immunol 11:2067. Frontiers Media S.A.
108. Kambouris ME, Manoussopoulos Y, Velegraki A, Patrinos GP (2023) The biote-bot hybrid. The ultimate biothreat merging nanobots, AI-enabled cybernetics and synthetic biology. Future Med AI 1:FMAI4 [Internet]. Future Medicine. Available from: https://doi.org/10.2217/fmai-2023-0008
109. Croom E (2012) Metabolism of xenobiotics of human environments. In: Hodgson E (ed) Progress in molecular biology and translational science. Academic Press, pp 31–88
110. Ascoët S, De Waard M (2020) Diagnostic and therapeutic value of aptamers in envenomation cases. Int J Mol Sci 21:3565. MDPI AG
111. Saito M, Uchida N, Furutani S, Murahashi M, Espulgar W, Nagatani N et al (2018) Field-deployable rapid multiple biosensing system for detection of chemical and biological warfare agents. Microsyst Nanoeng 4. Nature Publishing Group
112. Vines JB, Yoon JH, Ryu NE, Lim DJ, Park H (2019) Gold nanoparticles for photothermal cancer therapy. Front Chem 7 [Internet]. Frontiers Media SA. Available from: /pmc/articles/PMC6460051/. Cited 20 Aug 2022
113. Schulz K, Pöhlmann C, Dietrich R, Märtlbauer E, Elßner T (2019) An electrochemical fiveplex biochip assay based on anti-idiotypic antibodies for fast on-site detection of bioterrorism relevant low molecular weight toxins. Toxins (Basel) 11:696. MDPI AG
114. Schmidt M, Pei L, Budisa N (2018) Xenobiology: state-of-the-art, ethics, and philosophy of new-to-nature organisms. Adv Biochem Eng Biotechnol 162:301–316. Springer Science and Business Media Deutschland GmbH

115. Nagies FSP, Brueckner J, Tria FDK, Martin WF (2020) A spectrum of verticality across genes. PLOS Genet 16:e1009200 [Internet]. Public Library of Science. Available from: https://doi.org/10.1371/journal.pgen.1009200. Cited 23 Apr 2022
116. Liu G, Lin Q, Jin S, Gao C (2022) The CRISPR-Cas toolbox and gene editing technologies. Mol Cell 82:333–347 [Internet]. Available from: https://pubmed.ncbi.nlm.nih.gov/34968414/. Cited 5 Dec 2023
117. Shi J, Zhou T, Chen Q (2022) Exploring the expanding universe of small RNAs. Nat Cell Biol 24:415–423
118. Cech TR, Steitz JA (2014) The noncoding RNA revolution-trashing old rules to forge new ones. Cell 157:77–94 [Internet]. Available from: https://doi.org/10.1016/j.cell.2014.03.008. Cited 11 Apr 2024
119. Thai DA, Liu Y (2025) Nucleic acid amplification tests in digital microfluidics: the promise of next-generation point-of-care diagnostics. Microsyst Nanoeng 11:155 [Internet]. Available from: https://doi.org/10.1038/s41378-025-00977-5
120. Goudoudaki S, Kambouris ME, Siamoglou S, Gioula G, Kantzanou M, Manoussopoulou M et al (2023) Can water-only DNA extraction reduce the logistical footprint of biosurveillance and planetary health diagnostics? Toward a new method. OMICS 27:116–126 [Internet]. Available from: https://pubmed.ncbi.nlm.nih.gov/36809194/. Cited 27 Mar 2023
121. Goudoudaki S, Milioni A, Kritikou S, Velegraki A, Patrinos GP, Gioula G et al (2021) Fast, scalable, and practical: an alkaline DNA extraction pipeline for emergency microbiomics biosurveillance. Omics J Integr Biol 25:484–494 [Internet]. Mary Ann Liebert Inc. Available from: https://doi.org/10.1089/omi.2021.0090. Cited 6 Oct 2022
122. Wang X, Hong X-Z, Li Y-W, Li Y, Wang J, Chen P et al (2022) Microfluidics-based strategies for molecular diagnostics of infectious diseases. Mil Med Res 9:11 [Internet]. Available from: https://doi.org/10.1186/s40779-022-00374-3
123. Priye A, Wong S, Bi Y, Carpio M, Chang J, Coen M et al (2016) Lab-on-a-drone: toward pinpoint deployment of smartphone-enabled nucleic acid-based diagnostics for mobile health care. Anal Chem 88:4651–4660
124. Boguraev A-S, Christensen H, Bonneau A, Pezza J, Nichols N, Giraldez A et al (2017) Successful amplification of DNA aboard the International Space Station. npj Microgravity 3
125. Chávez-Ramos K, Trejo F, Salas-Navarrete PC, Ramón-Gallegos E, Muñoz-Medina JE, Alvarez-Icaza L et al (2025) Critical aspects of droplet digital reverse transcription loop-mediated isothermal amplification (ddRT-LAMP) for viral pathogens detection. Microsyst Nanoeng 11:137 [Internet]. Available from: https://doi.org/10.1038/s41378-025-00982-8
126. Christensen DR, Hartman LJ, Loveless BM, Frye MS, Shipley MA, Bridge DL et al (2006) Detection of biological threat agents by real-time PCR: comparison of assay performance on the R.A.P.I.D., the LightCycler, and the Smart Cycler platforms. Clin Chem 52:141–145 [Internet]. Available from: http://www.ncbi.nlm.nih.gov/pubmed/16391330. Cited 10 Jun 2019
127. Kambouris MEME (2016) Population screening for hemoglobinopathy profiling: is the development of a microarray worthwhile? Hemoglobin 40:240–246 [Internet]. Available from: https://doi.org/10.1080/03630269.2016.1186686. Cited 13 June 2019
128. Velegraki A (2020) Panmicrobial Microarrays. In: Kambouris ME, Velegraki A (eds) Microbiomics: dimensions, applications, and translational implications of human and environmental microbiome research, 1st edn. Elsevier Academic Press, London, pp 95–120
129. Wang H-Y, Luo M, Tereshchenko IV, Frikker DM, Cui X, Li JY et al (2005) A genotyping system capable of simultaneously analyzing >1000 single nucleotide polymorphisms in a haploid genome. Genome Res 15:276–283 [Internet]. Available from: https://doi.org/10.1101/gr.2885205. Cited 13 June 2019
130. Niedringhaus TP, Milanova D, Kerby MB, Snyder MP, Barron AE (2011) Landscape of next-generation sequencing technologies. Anal Chem 83:4327–4341 [Internet]. Available from: https://pubmed.ncbi.nlm.nih.gov/21612267/. Cited 8 Oct 2020
131. Patrinos GP, Pasparakis E, Koiliari E, Pereira AC, Hünemeier T, Pereira LV et al (2020) Roadmap for establishing large-scale genomic medicine initiatives in low- and middle-income

countries. Am J Hum Genet 107:589–595 [Internet]. Available from: https://pubmed.ncbi.nlm.nih.gov/33007198/. Cited 6 Feb 2023

132. Amer K, Soliman NA, Soror S, Gad YZ, Moustafa A, Elmonem MA et al (2025) Egypt genome: towards an African new genomic era. J Adv Res 71:415–427
133. Thermes C (2014) Ten years of next-generation sequencing technology. Trends Genet 30:418–426
134. Lee H, Gurtowski J, Yoo S, Nattestad M, Marcus S, Goodwin S et al (2016) Third-generation sequencing and the future of genomics online methods list of figures
135. Heather JM, Chain B (2016) The sequence of sequencers: the history of sequencing DNA. Genomics 107:1–8 [Internet]. Available from: http://www.ncbi.nlm.nih.gov/pubmed/26554401. Cited 11 Feb 2019
136. Gao F, Keinan A (2016) Explosive genetic evidence for explosive human population growth. Curr Opin Genet Dev 130–139. Elsevier Ltd.
137. Eizenga JM, Novak AM, Sibbesen JA, Heumos S, Ghaffaari A, Hickey G et al (2020) Pangenome graphs. Annu Rev Genomics Hum Genet 21:139–162 [Internet]. Available from: https://pubmed.ncbi.nlm.nih.gov/32453966/. Cited 7 Sept 2023
138. Berg G, Rybakova D, Fischer D, Cernava T, Vergès MCC, Charles T et al (2020) Microbiome definition re-visited: old concepts and new challenges. Microbiome 8:103. BioMed Central
139. Thomas T, Gilbert J, Meyer F (2014) Metagenomics: a guide from sampling to data analysis. Role Bioinform Agric 2:357–383 [Internet]. BioMed Central Ltd. Available from: http://www.microbialinformaticsj.com/content/2/1/3
140. Jager et al (2026) A breath of fresh air: a novel passive airborne eDNA approach for scalable terrestrial biodiversity monitoring. Ecol Evol. https://doi.org/10.1002/ece3.73346
141. Nelson WC, Tully BJ, Mobberley JM (2020) Biases in genome reconstruction from metagenomic data. PeerJ 8:e10119
142. Khalil AT, Shinwari ZK (2015) Threats of agricultural bioterrorism to an agro dependent economy; what should be done? J Bioterror Biodef 05:1–8 [Internet]. OMICS Publishing Group. Available from: https://www.omicsonline.org/open-access/threats-of-agricultural-bioterrorism-to-an-agro-dependent-economy-what-should-be-done.php?aid=27096. Cited 21 Sept 2020
143. Rosenbäck RG, Svensson A (2023) Resilience in keeping the balance between demand and capacity in the COVID-19 pandemic, a case study at a Swedish middle-sized hospital. BMC Health Serv Res 23:202 [Internet]. Available from: https://doi.org/10.1186/s12913-023-09182-4
144. Pujawan IN, Bah AU (2021) Supply chains under COVID-19 disruptions: literature review and research agenda. Supply Chain Forum 23:81–95 [Internet]. Taylor & Francis. Available from: https://doi.org/10.1080/16258312.2021.1932568. Cited 23 Aug 2022
145. Mazzucato M, Kattel R (2020) COVID-19 and public-sector capacity. Oxf Rev Econ Policy 36:S256–S269 [Internet]. Available from: https://doi.org/10.1093/oxrep/graa031
146. Fomsgaard A, Rosenstierne M (2020) An alternative workflow for molecular detection of SARS-CoV-2—escape from the NA extraction kit-shortage, Copenhagen, Denmark, Mar 2020. Eurosurveillance 25:2000398 [Internet]. European Centre for Disease Prevention and Control (ECDC). Available from: https://pubmed.ncbi.nlm.nih.gov/32290902/. Cited 21 Dec 2020
147. Read JM, Green CA, Harrison EM, Docherty AB, Funk S, Harrison J et al (2021) Hospital-acquired SARS-CoV-2 infection in the UK's first COVID-19 pandemic wave. Lancet 398:1037–1038 [Internet]. Elsevier B.V. Available from: http://www.thelancet.com/article/S0140673621017864/fulltext. Cited 7 Sept 2023
148. Sharma A, Farouk IA, Lal SK, Martinez-Sobrido L, Toral FA (2021) COVID-19: a review on the novel coronavirus disease evolution, transmission, detection, control and prevention. Viruses 13:202 [Internet]. Multidisciplinary Digital Publishing Institute. Available from: https://www.mdpi.com/1999-4915/13/2/202/htm. Cited 20 Apr 2022
149. Zhang D, Duran SSF, Lim WYS, Tan CKI, Cheong WCD, Suwardi A et al (2022) SARS-CoV-2 in wastewater: from detection to evaluation. Mater Today Adv 13:100211. Elsevier

150. Wee SK, Sivalingam SP, Yap EPH (2020) Rapid direct nucleic acid amplification test without RNA extraction for SARS-CoV-2 using a portable PCR thermocycler. Genes (Basel) 11:1–13 [Internet]. MDPI AG. Available from: https://pubmed.ncbi.nlm.nih.gov/32570810/. Cited 3 Nov 2020
151. Primmerman C (2000) Detection of biological agents. Lincoln Lab J 12:3–31
152. Kambouris ME, Siamoglou S, Kordou Z, Milioni A, Vassilakis S, Goudoudaki S et al (2020) Point-of-need molecular processing of biosamples using portable instrumentation to reduce turnaround time. Biosaf Health 2:177–182
153. Psarias G, Iliopoulou E, Liopetas I, Tsironi A, Spanos D, Tsikrika A et al (2020) Development of rapid pharmacogenomic testing assay in a mobile molecular biology laboratory (2MoBiL). Omics J Integr Biol 24:660–666. Mary Ann Liebert Inc.
154. Shrestha K, Kim S, Han J, Zhang M, Parajuli S, Cho G (2025) Infectious disease diagnostic device using rapid and efficient qPCR assays on a multi-target chip: idream-qPCR. Microsyst Nanoeng 11:143 [Internet]. Available from: https://doi.org/10.1038/s41378-025-00972-w
155. Ali N, Rampazzo RDCP, Costa ADIT, Krieger MA (2017) Current nucleic acid extraction methods and their implications to point-of-care diagnostics. Biomed Res Int [Internet]. Hindawi Limited. Available from: https://pubmed.ncbi.nlm.nih.gov/28785592/. Cited 10 May 2021
156. Howden BP, McEvoy CRE, Allen DL, Chua K, Gao W, Harrison PF et al (2011) Evolution of multidrug resistance during *Staphylococcus aureus* infection involves mutation of the essential two component regulator WalKR. PLoS Pathog 7 [Internet]. Available from: https://pubmed.ncbi.nlm.nih.gov/22102812/. Cited 19 Apr 2022
157. Perfumo A, Freiherr von Sass GJ, Nordmann E-L, Budisa N, Wagner D (2020) Discovery and characterization of a new cold-active protease from an extremophilic bacterium via comparative genome analysis and in vitro expression. Front Microbiol 11:881 [Internet]. Frontiers Media S.A. Available from: http://www.ncbi.nlm.nih.gov/pubmed/32528424. Cited 25 Sept 2020
158. Voelkerding KV, Dames SA, Durtschi JD (2009) Next-generation sequencing: from basic research to diagnostics. Clin Chem 55:641–658 [Internet]. Available from: http://www.ncbi.nlm.nih.gov/pubmed/19246620. Cited 15 Oct 2019
159. Gargis AS, Cherney B, Conley AB, Mclaughlin HP, Sue D (2019) Rapid detection of genetic engineering, structural variation, and antimicrobial resistance markers in bacterial biothreat pathogens by nanopore sequencing. Nat Sci Rep 9:13501 [Internet]. Available from: https://doi.org/10.1038/s41598-019-49700-1. Cited 1 July 2023
160. Palatnick A, Zhou B, Ghedin E, Schatz MC (2020) iGenomics: comprehensive DNA sequence analysis on your smartphone. Gigascience 9:1–12 [Internet]. Available from: http://orcid.org/0000-0002-4118-4446. Cited 15 Dec 2020
161. Ganguli A, Mostafa A, Berger J, Aydin MY, Sun F, Stewart de Ramirez SA et al (2020) Rapid isothermal amplification and portable detection system for SARS-CoV-2. Proc Natl Acad Sci U S A 117:22727–22735 [Internet]. National Academy of Sciences. Available from: https://pubmed.ncbi.nlm.nih.gov/32868442/. Cited 3 Nov 2020
162. Herrera S, Vallejo AF, Quintero JP, Arévalo-Herrera M, Cancino M, Ferro S (2014) Field evaluation of an automated RDT reader and data management device for *Plasmodium falciparum*/*Plasmodium vivax* malaria in endemic areas of Colombia. Malar J 13. BioMed Central Ltd.
163. Primiceri E, Chiriacò MS, Notarangelo FM, Crocamo A, Ardissino D, Cereda M et al (2018) Key enabling technologies for point-of-care diagnostics. Sensors (Switzerland) [Internet]. MDPI AG. Available from: https://pubmed.ncbi.nlm.nih.gov/30355989/. Cited 8 Oct 2020
164. Thang TD, Son DT (2025) Genetic testing in adult critical care across resource-limited Asian and African countries: barriers and pathways forward. Acad Mol Biol Genomics 2 [Internet]. Available from: https://www.academia.edu/3064-9765/2/3/10.20935/AcadMolBioGen7883
165. Kambouris ME, Reichard U, Legakis NJ, Velegraki A (1999) Sequences from the aspergillopepsin PEP gene of *Aspergillus fumigatus*: evidence on their use in selective PCR

identification of *Aspergillus* species in infected clinical samples. FEMS Immunol Med Microbiol 25:255–264 [Internet]. Available from: http://www.ncbi.nlm.nih.gov/pubmed/10459580. Cited 13 June 2019
166. Kubyshkin V, Budisa N (2017) Synthetic alienation of microbial organisms by using genetic code engineering: why and how? Biotechnol J 12. Wiley-VCH Verlag
167. Suliga W, Burnham RL, Deely T, Gavert W, Pronko MS, Verdun G et al (1999) Short-range biological standoff detection system (SR-BSDS). In: Air monitoring and detection of chemical and biological agents II. SPIE, p 72
168. Kambouris ME, Velegraki A, Patrinos GP, Poulas K (2014) Reviewing the US select agents and toxins list with more efficient assessment of the potential applicability of microbes in bioterrorism: the fungal agent. Am J Bioterror Biosecur Biodef 1:3–5 [Internet]. Available from: www.austinpublishinggroup.com. Cited 15 Oct 2019
169. Oliveira M, Mason-Buck G, Ballard D, Branicki W, Amorim A (2020) Biowarfare, bioterrorism and biocrime: a historical overview on microbial harmful applications. Forensic Sci Int 314:110366. Elsevier Ireland Ltd.
170. Huffman JA, Perring AE, Savage NJ, Clot B, Crouzy B, Tummon F et al (2020) Real-time sensing of bioaerosols: review and current perspectives. Aerosol Sci Technol 54:465–495 [Internet]. Taylor & Francis. Available from: https://doi.org/10.1080/02786826.2019.1664724
171. Kambouris M, Papathanasiou A-F (2021) Surveillance policies, levels, patterns and techniques. In: Genomics in biosecurity, 1st edn. Elsevier Academic Press, London, pp 43–60
172. Šválbenská E (2012) Systems for detection and identification of biological aerosols. Def Sci J 62:404–411 [Internet]. Defense Scientific Information and Documentation Centre. Available from: https://publications.drdo.gov.in/ojs/index.php/dsj/article/view/1251/4619. Cited 2 Nov 2020
173. Savage NJ, Krentz CE, Könemann T, Han TT, Mainelis G, Pöhlker C et al (2017) Systematic characterization and fluorescence threshold strategies for the wideband integrated bioaerosol sensor (WIBS) using size-resolved biological and interfering particles. Atmos Meas Technol 10:4279–4302 [Internet]. Available from: https://amt.copernicus.org/articles/10/4279/2017/
174. Thrush E, Salciccioli N, Brown DM, Siegrist K, Brown AM, Thomas ME et al (2012) Backscatter signatures of biological aerosols in the infrared. Appl Opt 51:1836–1842 [Internet]. Optica Publishing Group. Available from: https://opg.optica.org/ao/abstract.cfm?URI=ao-51-12-1836
175. Maidment L, Zhang Z, Howle CR, Reid DT (2016) Stand-off identification of aerosols using mid-infrared backscattering Fourier-transform spectroscopy. Opt Lett 41:2266–2269 [Internet]. Optica Publishing Group. Available from: https://opg.optica.org/ol/abstract.cfm?URI=ol-41-10-2266
176. Ifarraguerri A, Ben-David A, Vanderbeek R (2009) Estimating the limit of bio-aerosol detection with passive infrared spectroscopy. In: Spectral sensing research for surface and air monitoring in chemical, biological and radiological defense and security applications. World Scientific, pp 475–485 [Internet]. Available from: https://doi.org/10.1142/9789812835925_0037
177. Joshi D, Kumar D, Maini AK, Sharma RC (2013) Detection of biological warfare agents using ultra violet-laser induced fluorescence LIDAR. Spectrochim Acta Part A Mol Biomol Spectrosc 446–456 [Internet]. Elsevier B.V. Available from: https://pubmed.ncbi.nlm.nih.gov/23719340/. Cited 29 Sept 2020
178. Liu Q, Yan S, Zhang M, Wang C, Xing D (2024) Air sampling and ATP bioluminescence for quantitative detection of airborne microbes. Talanta 274:126025 [Internet]. Available from: https://www.sciencedirect.com/science/article/pii/S0039914024004041
179. Wilcoxson C, Shand SM, Shand RF (1999) Kitchen microbiology (it's easier than you think!). Am Biol Teach 61:34–38 [Internet]. Available from: https://doi.org/10.2307/4450606

Part II
The Meta-Biothreat

Chapter 3
Exotic Biothreats

Abstract In the context of the Meta-Biothreat new breeds of pathogens may be used to derail surveillance and intervention. Such cases are the fully or partly engineered strains—Metapathogens but also the Neopathogens. The latter are species and strains emerging or discovered recently and not yet properly processed by surveillance means and concepts, nor by the adaptive collective immunity or the medical intervention establishment. Such entities may be found in secluded locations on earth, in what may be termed "biospheric bubble", or in remote areas within the biosphere but isolated due to ecologic or geographic/geophysical barriers and they may be of similar to existing ones and studied by Microbiology, or of totally different biochemical basis, being in the realm of Xenobiology. Similarly, Astrobiology tackles entities found in space, similar to terrestrial ones if the *Panspermia* theory holds water and studied by Exobiology; or totally different and studied once more by Xenobiology. Other cases of exotic microorganisms may be unclassified, but the biobots refer to existing cells combined to higher structures hitherto unknown and thus forming new organisms, while de-extinction recreates extinct and thus unknown but fully evolved microorganisms, as happened with the Spanish flu virus.

Keywords Exobiology · Xenobiology · Astrobiology · Gigantic viruses · Biobots · Extremophilia · Re-wilding efforts · Radiosynthetic organisms · De-extinction · *Panspermia* theory

The somewhat lyric and catchy title of this chapter refers to the very heterogeneous and extremely well-unknown cases of bioagents that may be used so as to circumvent surveillance, resilience and possible response but also to alleviate hiccups and pitfalls in the prospective perpetrators' efforts to acquire usable (not necessarily weaponizable) bioagents. A detailed mapping of such elements and cases in imperative in order to thwart their use and implementation within the concept of Disruptive Biosecurity. It is a task for dedicated, permanent and possibly multidisciplinary study groups of the concerned stakeholders.

M. E. Kambouris, *Preparing for the Meta-Biothreat: Disruptive Biosecurity*, Advanced Sciences and Technologies for Security Applications,
https://doi.org/10.1007/978-3-032-25690-4_3

3.1 Seeking the Unexpected

The two most important traits in selecting a bioagent, from the perpetrator's point of view, are (i) the resistance to direct countermeasures, and (ii) the evasion of detection and identification—in essence of the agnostic or targeted surveillance and of the (always targeted) screening efforts. The (i) incorporates first and foremost the realm of immune response, both spontaneous (natural immune capacity and response) and artificially leveraged. The first spans across both natural immunological mechanisms, Innate and Adaptive, where applicable [1]. The second includes the infusion of antisera, the immunization by vaccination, plus the use of any therapeutic amenities based on immunity-related molecules and molecular fragments. For toxins and bioregulators, the use of antidotes may be included in this category, acting on the detrimental amenity directly, the example being Sodium thiosulfate that transforms toxic cyanide residues to stable, benign derivatives [2], or may repair the damage caused and is thus to be counted among the drugs (see below), as in oxime treatment of organophosphoric poisonings, as with nerve agents [3].

Then comes the realm of the drugs, acting directly on the bioagent (antibiotics) [4] or alleviating its detrimental effects, as with anti-inflammatory drugs to interrupt any cytokine storm [5]. The latter is the field of the supportive care/symptomatic therapy, which may be considered causative therapy regarding the suppression of secondary, downstream damage. The former consists causal therapy as it removes the causal agent, but may still miss some serious secondary/collateral damage induced latently as in infection-induced/facilitated tumorigenesis [6] or directly downstream, as with the cytokine storm [7].

The "drugs" are in this book to be understood as all amenities of chemical/biochemical nature, irrespective of their source of production, which act upon the bioagent or the host organism so as to weaken, neutralize or utterly destroy the former or to empower and restore the latter; it does not include the serological approaches, with antisera and monoclinic antibodies as mentioned above. Also not included are the disinfectants/decontaminants/sterilizers that are too potent for internal, *in vivo* use and are employed to incapacitate/destroy bioagents before their entry into the organism and the biochemical environment of their intended hosts; at this point the pathogens may be more robust and survivable, especially if they are dispersed in some latent form, but even more powerful and drastic amenities may be used to overcome the excess survivability of the dispersal forms of the pathogens. This early, external intervention thus constitutes a very promising approach for the application of countermeasures, and the only one suitable for long-prospect reclaiming of polluted/contaminated/infected space and objects [4, 8].

A sub-category of the wider "drugs" category of therapeutic and prophylactic amenities are the Biopharmaceuticals. Contrary to the usual (chemo)pharmaceuticals which are mainly (but not exclusively) micromolecular drugs produced by chemical procedures and respective equipment, the biopharmaceuticals, or "biological", are produced by living organisms or *are* living (micro)organisms [9, 10]. They include but are not limited to the antimicrobial peptides, which are molecules, but much bigger than the usual molecules of micromolecular antibiotics—irrespective if the latter are produced by chemical synthesis or as a copy and modification of a natural antimicrobial molecule. And then there are whole organisms, such as the viruses for virotherapy against bacteria [11, 12] but possibly also against Eucarya, such as fungi, protozoan and metazoan parasites [13]; an approach rudimentarily described as "an infection of the infection". Similar is the prospective use of both viruses and bacteria to destroy cancerous cells in tumors [14, 15], which, by their live status actually feed on the target cancerous cells and multiply, a process actually amplifying the dosage used; it is in effect "an infection against cancer" by "oncolytic viruses" [14] and "oncolytic bacteria" [16].

As obvious from the above, bioagents that may endure these direct pathways of their incapacitation (external and environmental disinfection/decontamination, direct and indirect immunization, systematic chemotherapy and systematic biotherapy) are extremely hard to counter with decontamination and medical countermeasures. Thus such traits are highly appreciated within the dark world of bio-offense [17, 18]. On the other hand, if an agent is not detected, as with trait (ii) above, even if it is highly susceptible to all the above (but for the natural immune capacity of the target, Innate and Adaptive) it is relatively survivable: not detecting and identifying it, negates the use of any prophylaxis or empiric therapy in the first case, or targeted therapy in the second case. Only symptomatic/supportive treatment may be administered. This course by itself is laborious and costly, as in the case of Intensive Care Unit-ICU admittance [19], which, to a considerable extend might be a satisfactory disruptive effect, fully justifying the method and the act of the bio-offense. It becomes obvious that engineering agents, or finding novel, hitherto unknown agents (metapathogens/neopathogens [20]) that are stealth to the current surveillance/screening/testing amenities may be a key enabler for the bioperpetrator, as a concerted effort to test the agents' potential to shirk detection or, at least, identification, is much easier than to test their resistance and survivability against antimicrobial amenities, especially drugs functioning within the target organism and assisted by its immunity. The former needs some diagnostics/detection instrumentation and basic skillsets, the latter actual experiments with living organisms and the respective competencies, leaving a pronounced trail that authorities may pick up and track.

3.2 Estranged Cousins and Outright Aliens

As a basic rule, in Biology and Biosciences in general, the prefix "auto-" (Greek for "self-") is used for entities produced within an individual organism or very close relatives, ideally but not necessarily clones. The prefix "allo-", Greek for "other" is used for similar entities produced or coming from an organism of the same species, but not the reference organism, for example a transplant, i.e. kidney or liver ("allograft"). And the prefix "xeno-" (Greek for "foreign-er") is used for anything else, more distant (which is a very wide and imprecise a prerogative); from another species, as in some grafts, or coming from another world, might that be alien, or not.

As a result, there is some controversy on the use of the term "Xenobiology" and the respective use of the prefix "xeno-" to any other, subjugate entities, the main example being "xeno(micro)organisms" [21] but also "xenobiotics" [22]. The term may very well apply to products of synthetic biology (the subject of Chap. 4) which are *different*, in that they use molecules different of and possibly incompatible with the ones of the earthen biochemistry, in any of a number of levels; the main being the DNA, where major chemical differences create XNA (Xenobiotic Nucleic Acids) instead, and of many different parallel iterations [23–25]. But the term may very well be used for *alien* (micro)organisms coming from another planet, solar system or galaxy, which is actually the realm of Astrobiology [26–29] with the diverse definitions of Life that may or may not apply [30]. Such lifeforms may be of a very different biochemistry, an example being a Silicon Oxide-SiO biochemical basis instead of the Carbon-C basis of the terrestrial one [31]. Their physiology would be dissimilar or similar, but there is the prerogative of fulfilling the characteristics of Life as set in Earth, or slightly expanded or otherwise differentiated. Such (micro)organisms, if of a drastically different physiology, may not interact with terrestrial lifeforms, or may interact in a limited way, which, however might be extremely dangerous, should they be able to use earthen amenities for energy and matter uptake; or not, if they cannot. Essentially different biochemistries may not be functional or survivable in earthen biosphere conditions and thus the biothreat might be limited or null, at least conditionally. However, within technidal environments, like ice or other refrigeration conditions, vacuum and various sets of conditions in storehouses and storage or processing facilities, such as tanks of fuel or tanks and reactors of the chemical industry, they may find niches of survival and exercise a more or less detrimental effect against technidal and other inorganic amenities, especially high-cost, low-abundance and high-importance ones, a hypothetical example being the fiber-optics cables.

Or such organisms from galaxies far, far away (or planets, meteorites and any other conceivable stellar body), might be more like earthen microbiota in terms of elementary constitution, if there is any merit in the *Panspermia* theory [32, 33]. This is actually the field of the Exobiology; to question the existence of earth-like Life in other stellar bodies [28, 29], while Astrobiology engages in the study of stellar Life irrespectively of likeness to earthen Biosphere.

The case of earthen organisms isolated from the Biosphere by seclusion due to some absolute nature of any kind of barriers [34] would share similar principle with Astro-agents. They might have similar biochemistry to the Biospheric dwellers, but different evolution/pedigree stemming from *some* ancestor and appearing downstream. But they may be of entirely different biochemistries: the last case may be as diverse as the existence of Si–O backbone instead of C–C [31]; or less so, as the shift in the actual biochemistry being due to evolutionary pressure, as in the substitution of Phosphorus-P with Arsenic-As in nucleic acids [35]. Or less divergent, as the—already observed—appearance of very radical adaptations within the context of conventional biochemistry, such as the radiosynthetic mechanisms of fungi in deactivated but highly radioactive nuclear plants [36], and the chemotrophic pathways in rock-embedded microbiota [37].

In any case, the existing, standard antimicrobial amenities are not expected to show activity against xenomicrobiota that use a very different physiology [21]. Still, with the possible exception of meteorite impacts, it is the least probable option as space travel to inhabited worlds (understood as housing Life) will not happen for decades to come, although chains of events might cause the improbable, either directly with some impact to the biosphere (not necessarily all the way to the surface-the ground or the sea) or indirectly, through returning elements of space exploration/exploitation. Much more probable is the possibility of such returning elements to bring back earthen microbiota, exposed and thus mutated in extremely adverse conditions, and thus possibly very aggressive and/or resistant to antimicrobial amenities, especially irradiation. Such microbiota have been detected and analyzed already: one case comes from the International Space Station-ISS whence the NASA recovered a drastically differentiated and evolved strain with even greater resistance than the original, multi-drug resistant Gram-negative *Enterobacter bugandensis.* The *E. bugadensis* is a relatively novel pathogen which causes neonatal sepsis and severe hospital infections and it distributes multi-drug resistance within a given microbiome by plasmid transfer [38, 39]. Another case comes from the Chinese space mission Shenzhou-15. A bacterium, *Nialla tiangongensis* [40] came back from the space station Tian-gong (as indicated in name of the new isolated species). It is a relative of the *Nialla circulans*, a Gram-positive spore-producing rod-like bacterium with the ability to hydrolyze gelatin for acquiring nitrogen and carbon. *Nialla tiangongensis* demonstrates biofilm formation and possesses great resilience to oxidative stress, due to an elaborate oxidative stress response, and potentially to radioactive toxicosis as well, as it also possesses a very efficient radiation damage repair mechanism, thereby being able to survive in the extremely adverse space environment.

Thus, taking earthen microbiota for a space excursion, possibly as indoor/environmental contaminants or as parts of the human microbiomes of the crews, may result in some greatly amplified mutants-Neopathogens. Given that the precautions against space bacteria (of any kind and origin) are rather inadequate [26, 41], not only space-originating epidemics are possible, but interested parties with access to the spacefare returns, private or public might these be, have direct and perhaps indirect access to such strains. It is easy to understand that the robustness against the space environment would make them immune to a many anti-microbial amenities, either directly as a result of their adaptation, as is the UV and extremely low temperatures, or indirectly so: adaptations aimed for the above two factors may provide resistance to phagocytosis or chemical penetration/permeation.

Another source of unknown microbiota, with increased possibilities to have traits of undetectability or different, mainly increased, pathogenicity stems from secluded environments on Earth, where ancient strains have evolved along different pedigrees than cousin microorganisms—"cousin" implying that they share an ancient ancestor. One could be so bold as to suggest that the causality between the two prokaryotic Domains, the Archaea and the Bacteria might have been such an isolation, leading to the divergent pedigree of the latter and the Eucarya [42, 43]. These environments may be isolated due to some geophysical, geographical or ecological barrier, which severed exchanges of genetic information and possibly—but not necessarily—other matter with the Biosphere. The conditions though may be similar, more or less, to certain areas of the Biosphere, thus resulting in organisms with some common qualities, as are the environmental temperature tolerance, acidity, humidity or the pH and the basic nutritional needs. Such may be, indicatively and not exhaustively: the hidden valleys and tropical mega-rivers [44], marine environments isolated by streams or thermocline layers [45], simply deep soil layers [34], cave systems sealed by avalanches [46], frozen habitats from where the giant Alphapithovirus has been retrieved and resurrected [47], subterranean hollows sealed off by ice caps [48, 49], or even surface lakes fully enclosed by large swaths of deserts and with no connection with underground water systems. These areas may well house microbiota not co-evolved with biosphere macroorganisms colonized or parasitized by native microflora and thus the latter may be extremely susceptible to the former, having developed no immunity (Innate or Adaptive) whatsoever against them. The imbalance may work in the opposite sense; the species of the secluded environment may be preyed upon or antagonized by current biospheric species to the formers' extinction.

Secluded environments may be, on the other hand, functionally, not physically isolated. Great altitudes and depths, extremely high pressures and temperatures or extremely low ones and a lack of essential nutrients and chemical elements may spur evolution of drastically different lifeforms. The extremophilia first comes to mind [50] and would result in organisms highly resistant either to extreme conditions encountered in some methods of deployment, like the pressure and heat of explosive deployment by shells and bombs, or, more importantly, the extreme conditions used for sterilization/decontamination [51]. These include some corrosive chemicals like the free radicals/reactive oxygen species-ROS [52], the high temperatures, sudden high doses of X-ray, microwave, radioactivity or UV irradiation [53–55] and high pressure [56]. In some cases secretions that allowed survival in extreme environments may be toxic for species of more temperate ecosystems, but, which is very important, such secretions may be detrimental to materiel: enzymes of Actinobacteria degrade rubber [57] and enzymes evolved to crack rock or marble [37] would easily destroy the integrity of expensive linings for stealth, environmental protection or spectrum filtering, as in electro-optics. High extremophilia might allow space deployment of such microbiota to degrade satellite equipment, especially telecommunications and surveillance sensors. There is some controversy over microbiota substituting Phosphorus/P with Arsenic/As [35] but the question remains: if all other elements remain "normal", as it seems, could this change be accommodated by the enzymes used for digestion and polymerization of normal, P-containing Poly-/Oligo-/Mononucleotides? If not, NAAT diagnosis is off, spectrometry also, but microscopy and serological approaches might remain relevant [58].

3.3 Enter the Undead

The *mantra* of conserving biodiversity sounds a bit weird; simply by being sponsored by some extremely high-end stakeholders, if for no other reason. Species become extinct regularly and there is no objective assessment of such rate, in terms of comparison: it has not been properly stated how many species were extinct along with the dinosaurs and neither the Planet nor Life suffered a glitch. Thus, despite some ecological reverses, there is actually no reason for more species not becoming extinct and being protected as "endangered". But in terms of economy it does perfect sense: each species has evolved, honed and fine-tuned in the wake of very long timeframes, a vast number of effector molecules, which is a treasure cove for industrial purposes and not only for the Health-related industry: From proteins, especially enzymes, to a wide array of ribozymes, polysaccharides, lipids and micromolecular compounds, such as antibiotics or antidotes [59] and diverse kinds of macromolecular ones, suitable for therapeutic reasons—the very concept of the biologicals/biopharmaceuticals [9, 10]. The development of processes, such as the Manta ray navigation [60] and the long-range whales' communication [61], but most of all of compounds, such as the spider silk which may be used for body armor [62, 63] are also of extreme interest for all kinds of industries and services; actually, for many, diverse and influential stakeholders.

3.3.1 The Eurydice Projects

As a result, the reintroduction of extinct species is much more than an ecological and evolutionary prodigy, the latter exemplified by the "re-wilding", i.e. introducing extinct species anew in their native habitats [64, 65]. This effort to reinstate the extinct, long dead and perhaps fossilized or refrigerated organism (hibernation and artificial refrigeration are not included in the concept) to living/extant status ("de-extinction" [66–68]) is reminiscent of the legendary effort of the divine musician Orpheus to bring his recently deceased love, Eurydice, back from the Underworlds. The actual recreation of extinct species by genetic relics, which may include some degree of repair/restoration or completion of damaged/missing Nucleic Acid sequences implies a degree of guesswork [68] as succinctly predicted in the Jurassic Park movies franchise. This actually is tantamount to the creation of a completely novel species, but irrespective of this dimension, which may lead to profound unpredictable interactions, and seems perfect for ecoterrorism, the idea of resurrecting extinct organisms, especially microorganisms, is highly disquieting *per se.* The return of extremely lethal pathogens to which no one has immunity as they are extinct and thus out of the (literal) living memory of immunization, like, for example, has already happened with the Spanish Flu [69], is a nightmarish prospect. The basic implementation of this approach is the production of unknown pathogens—or biochemical effector amenities [67, 70]—which have not been included in diagnostic manuals, training syllabuses and treatment protocols. So, they would present to any attending personnel (not only physicians) with an unforeseen riddle. The initial mortality and morbidity of any successful deployment of such agents would be catastrophic. Initially, only syndromic surveillance and symptomatic treatment would be of some predictive and therapeutic/prophylactic value respectively, until the agent is detected and—an issue sometimes escaping attendance—correctly identified, as it does not actually exist and no-one knows what to look for: nature, morphology, size, biomarkers, symptoms and signs.

In the most modern iteration, such revived organisms, even if not totally viable, may produce their genomic blueprints for inclusion in applications of synthetic biology regarding sequences, extracts, products. It is not necessarily their entirety, not their cellular or particle phase that may be exploited; or, at least, not only. Their informational or biochemical peculiarities, as in sequences or pathways is of essence. These revived organisms may not be survivable, robust and weaponizable or directly applicable in assault schemes, as the weapon potential of a microbe is a multi-parametric issue [71–74]. Actually, this may be a preferred condition, for reasons of security and safety, and synthetic auxotrophs are regularly considered a safe version of a dangerous strain to work with; Biocontainment is an essential feature when deploying genetically modified organisms (GMOs) so they may not escape, not be smuggled without the necessary supplements of its nutrient [75], and for non-nutrient dependent organisms, such as are the viruses, the key may be some custom-designed conservative.

The abovementioned subcellular/sub-particle entities may be used for enhancing some other existing microorganisms or included in the design of completely new artificial ones that would be designed from scratch. And perhaps the worst dimension of the problem would be that the specifics of any such multi-level and multi-aspect pathogenic amenities would be known only to the subject that revived/reconstituted the extinct source (micro)organism. Anyone else would not even suspect the (specifics of the) fact.

3.3.2 *"So the line may live on": the biobots*

A focal point in this book is the proposition (detailed in Chap. 8) that the interactions are just as important as the physical entities. As a result of such perspective, one may proceed a bit further from the conventional wisdom: an organism is the sum of its cells, but losing some of them to spontaneous decay or damage of any kind does not change the organism in essence, although it does change its composition cell-wise. Even massive losses, for example of a limp, do not alter the organism as such. Much more so if to recompense, prosthetics or transplants (the latter being similar cells but of another organism and genome and thus biomarker composition—a familiar alien of sorts) are employed. At the end, one may conclude, by repetitive, successive reduction paces, that it is the conscience and cognition, not the cells, the defining quality of the identity.

The interesting thing is what happens the other way around. The cells that are severed and viable for some time after separation (this time varies and may be set by consensus in a case by case basis), do they carry part of the identity? They cannot communicate and thus we cannot know how they feel or self-define. If some cells are not expected to enact such conscience, but may carry it, it would be of extreme if not existential interest to know what happens if they multiply in a permissive environment—culture media included but being only one option. Over what count the inherited identity will reveal and allow self-recognition of the community and self-consciousness? There is the possibility of course that if the cells simply proliferate and aggregate this might not happen. But if the progenitor of this population is not differentiated, or loses its differentiation somehow, will it and its spawn repeat the same ontogenetic procedure, step by step? And the end construct, will it carry the same identity and conscience, being a replicate being (not just an organism)?

These are questions of science and science fiction, and repeatedly movies and books have engaged in supposing, forecasting and projecting one way or, usually, the other, the more radical one. But in here there is a biological issue: what will be the result if the severed cells, especially if they are differentiated, start or continue proliferation, free of any limitations imposed by the multicellular environment, regarding position, attachments, (bio)chemical environment, nutrients supply and servicing, the last referring especially to waste disposal, regulation and defense/immunity. Usually, one tends to think of such eventualities in terms of undifferentiated cells; but if differentiated cells lose all regulation and control by their organismic environment and are in a favorable natural microenvironment, perhaps a simple aggregation would not be the end result; this aggregation may establish novel interactions and relations, producing a shaped, multi-cellular lifeform very different than the original one, or the ones known so far. This is the concept of the biobot (see Chap. 4), the name implying a cybernetic entity of biological, mostly organic and flexible form, intended to accomplish relatively simple tasks in environments and conditions suiting its characteristics [76].

These lifeforms are enticing interest to how they may be exploited, i.e. guided or trained to self-sufficiently overcome obstacles and adapt to the environment in order to execute a simple mission [77]. For example, as the same biomarkers are expected due to the clonal relation with the genome of the progenitor, such bots would not trigger immune responses and could deliver medication to targeted positions, entities and biocompartments to which they would have an affinity as per their pedigree, as severed but differentiated cells. And this delivery, instead of medicament or nutrient, could well include a toxin, a bioregulator, or a biohacking agent. Given that the genome is identical and the form unknown, it would be out of pure lack to detect such constructs by any autopsies, while regular surveillance, even in terms of personalized diagnostics, could not discriminate it against the genetically identical environment of the target organism.

Industrial and meta-industrial applications obviously abound, as in motility to carry solvents or disinfectants to microenvironments, including but not limited to microfluidics that may have been corrupted by excessive or improper use, thus becoming micro-plumbers. In the context of decentralized productive units of sustainable size and based mostly on biological procedures than conventional mechanical and chemical ones, as in Industry 4.0 and 5.0 [78–81], such amenities would be instrumental. And for military options, too. From thus servicing entities of military interest, to compromising similar ones of a target entity, in military, terrorist or criminal applications.

3.4 *Sui generis*: Being Unclassified

Marginal organisms not within established study boundaries, such as *Pneumocystis*, which wobbles between fungal and protozoan nature [82] and the viroids, which are molecules and are not considered possible animal pathogens [83] but are serious threats of food security [84–86], cause controversy and thus possible lapses in surveillance systems based on identification by taxonomy. It must be remembered that in days of old, the nature of Cyanobacteria, which are not human pathogens but occasionally produce Cyanotoxins (highly pathogenic as swimming water pollutants and foodborne contaminants [87]), were the subject of intense dispute. This revolved around the question of them being algae and thus plants, as they have internal membranes, contrary to bacteria, or a group within the kingdom *Monera*, as they have no secluded, defined nucleus with DNA organized in chromosomes [87, 88]. Then, the *Archaea* were also very controversial as they were drastically promoted from one group of the Kingdom *Monera*, the *Archaeobacteria* [88], to a Domain of Life [43].

The marginal organisms readily (may) possess characters in-between two taxa or traits that cause unforeseen combinations of diagnostic challenges and detrimental effects, throwing off targeted surveillance and occasionally syndromic surveillance, too. A novel such case is an organism between bacteria and viruses, the *Sukunaarchaeum mirabile.* It is an excessively reduced *Archaeon*, which may produce autonomously the core molecules for its replication, but its drastically truncated genome, only half the size of the next smaller cellular genome, results in an unprecedented dependence on its host, to which it delegates a very important part of its necessary functions [89]. This means that a similarly structured pathogen of another species, or an engineered strain of the original *S. mirabile* augmented with pathogenic traits by genetic manipulation [90], would not be affected by antibiotics targeting any of the outsourced functions.

The Mimivirus [91, 92] and the Klothoviruses [93] take the opposite direction. They are DNA viruses of extreme size; the latter is the biggest known virus, bigger than some bacteria, at perhaps 3μm in length. Both form extremely complicated virions, with multiple membranes within the capsid, and RNA molecules prepackaged and ready for immediate use in the host, thus bypassing the need to hijack the cellular host mechanism in order to produce mRNA, a somewhat tricky proposal as cell defenses focus heavily on not allowing alien Nucleic Acid transcription, for example with the small interference RNAs-siRNAs [94] but other options do exist

[95]. This implies some resilience against a number of antiviral amenities should there be, or be manufactured, strains pathogenic for species of interest, including but not limited to humans.

There is also a similar question regarding the evolutionary past of the poxviruses, where the *Variola major* virus, the causative agent of Smallpox, belongs. It is not clear whether viruses in general are a product of evolution towards higher complexity, perhaps starting from the viroids and adding features, improvements and capabilities (not to mention complexity and size). This would be a fitting idea for RNA viruses and a course similar to the one projected for the History of Life [96, 97]. But another possibility is for the viruses to be a product of devolution, cells (or latent forms of cells) that lost function and finally were stripped down to the bare essentials: the assisted reproduction. This would be a suitable interpretation for the creation of the DNA viruses, especially large and complex ones like the Mimivirus and the Klothovirus, as mentioned above, which share some features of structural complexity and functional peculiarities with the poxviruses, such as the transportation of already synthesized pre-packaged macromolecules to be delivered to the host upon infection. This devolution suggestion is bolstered by the discovery of intermediate entities, like the abovementioned *Sukunaarchaeum mirabile*, which have a cellular, not viral, origin [89]. Of course, both suggestions may be valid, and then some; viruses are so heterogeneous a Domain [98], that cannot be considered of a unique pedigree or of a single progenitor.

All the poxviruses are very large and complex entities, with the virion accompanied by two massive packages of useful proteins, the lateral bodies; a unique arrangement [99, 100]. The satellite viruses, which need another virus so as to act as one virus by themselves, are verging between Life and non-living entities [101, 102] but constitute a perfect example of binary systems (see Chap. 4). Viruses, after all, only recently were securely considered proper, although acellular, lifeforms and recognised as another, fourth Domain of Life [98] (or an even higher taxon, the Viral Empire, separate of the Empire of the cellular lifeforms [97, 103, 104]). This was done once they were found to be themselves preyed upon by some pathogens, the virophages [105]. The usefulness of virophages against pathogenic viruses is obvious [12, 106], a feature possibly applicable in Biosecurity, regarding preparations against viral pathogens of perpetrated deployment. The same applies to bacteriophages against pathogenic bacteria [11] and to other cases of hostile relations between/among microbiota; after all, the advent of antibiotics is exactly the result of this bellicose trait amongst the microbes in general [4].

Still, viruses with non-human/plant/animal virulence may also have a hidden threat potential that makes them excellent assault bioagents: they may destabilize health by attacking natural (appendage or environmental) microbiomes, which is obvious but not usually stated. The possible mechanics of such an offense are extremely complicated and far-fetching [107]. Furthermore, as the biochemicals and simple biologicals, such as in phage therapy and bacteriotherapy, acquire more traction [108, 109], the use of bacteriophages and virophages may well derail such therapies and thus constitute an indirect bioagent, not causing disease but undermining the medical efforts of intervention to alleviate or even cure other diseases. Actually, if the microbiome is seen as another tissue or organ of the host organism (a view widely but not universally embraced) [110], it is a logical approach to compromise or delete it, despite being operationally limited by its long-term effects rather than short-term efficiency.

And then there are the gigantic bacteria. Exemplified by the *Thiomargarita magnifica* [111], such bacteria, if modified in terms of biochemistry and engineered to acquire virulence—or if they are used in some hitherto hidden pathogenic application—would be quite a challenge. One issue is the expected septic/toxic effect (see Chap.7), amplified by the sheer size: just one cell would secrete large quantities of any innate or inserted/pasted toxin and digest large amounts of host tissue. The large size would also increase the interacting surface with the host and thus may increase the latter's inflammatory response. Furthermore, a possibly overlooked aspect is that the extremely large size may cause, even after few binary fission cycles, mechanical injury and disruption to host tissues and organs, a kind of damage a normal-sized pathogen would never be expected to achieve. Finally, the multiple copies of its genome would thwart quantitative NAAT diagnostic approaches targeting unique bacterial sequences, but also ubiquitous ones [58].

The very large size offers increased survivability: it would require very large quantities of antibiotics to achieve any pretense of neutralization (irrelevant whether stasis or necrosis [112]), thus limiting the range of usable compounds that may achieve the correct balance of pharmacodynamics versus pharmacokinetics [113], and of efficacy versus possible adverse reactions due to high dosage [114]. In the case of animal and human hosts, which are endowed with a complex immune system, the very large size would deplete many immune response amenities and makes the mega-cell utterly resistant to phagocytosis [115].

References

1. Xing Z, Afkhami S, Bavananthasivam J, Fritz DK, D'Agostino MR, Vaseghi-Shanjani M et al (2020) Innate immune memory of tissue-resident macrophages and trained innate immunity: re-vamping vaccine concept and strategies. J Leukoc Biol 108:825–834 [Internet]. Available from: https://pubmed.ncbi.nlm.nih.gov/32125045/. Cited 19 Apr 2022
2. Mégarbane B, Delahaye A, Goldgran-Tolédano D, Baud FJ (2003) Antidotal treatment of cyanide poisoning. J Chin Med Assoc 66:193–203
3. Kassa J (2002) Review of oximes in the antidotal treatment of poisoning by organophosphorus nerve agents. J Toxicol Clin Toxicol 40:803–816
4. Maris P (1995) Modes of action of disinfectants. Rev Sci Tech Off Int Epiz
5. Hu B, Huang S, Yin L (2021) The cytokine storm and COVID-19. J Med Virol 93:250–256 [Internet]. Available from: https://pubmed.ncbi.nlm.nih.gov/32592501/. Cited 20 Aug 2022
6. Travaglione S, Carlini F, Maroccia Z, Fabbri A (2024) Special issue "bacterial toxins and cancer". Int J Mol Sci
7. Chinta S, Rodriguez-Guerra M, Shaban M, Pandey N, Jaquez-Duran M, Vittorio TJ (2023) COVID-19 therapy and vaccination: a clinical narrative review. Drugs Context 12
8. Boyce JM (2016) Modern technologies for improving cleaning and disinfection of environmental surfaces in hospitals. Antimicrob Resist Infect Control 5:1–10

9. Kesik-Brodacka M (2018) Progress in biopharmaceutical development. Biotechnol Appl Biochem 65:306 [Internet]. Wiley-Blackwell. Available from: /pmc/articles/PMC6749944/ . Cited 23 Apr 2022
10. Declerck PJ (2012) Biologicals and biosimilars: a review of the science and its implications. Generics Biosimilars Initiat J 1:13–16
11. Adhya S, Merril CR, Biswas B (2014) Therapeutic and prophylactic applications of bacteriophage components in modern medicine. Cold Spring Harb Perspect Med 4 [Internet]. Cold Spring Harbor Laboratory Press. Available from: /pmc/articles/PMC3869276/ ?report=abstract. Cited 28 Sept 2020
12. Lin DM, Koskella B, Lin HC (2017) Phage therapy: an alternative to antibiotics in the age of multi-drug resistance. World J Gastrointest Pharmacol Ther 8:162 [Internet]. Baishideng Publishing Group Inc. Available from: https://pubmed.ncbi.nlm.nih.gov/28828194/. Cited 2 Mar 2021
13. Barrow P, Dujardin JC, Fasel N, Greenwood AD, Osterrieder K, Lomonossoff G et al (2020) Viruses of protozoan parasites and viral therapy: is the time now right? Virol J 17 [Internet]. Available from: https://pubmed.ncbi.nlm.nih.gov/32993724/. Cited 20 Aug 2022
14. Kaufman HL, Kohlhapp FJ, Zloza A (2015) Oncolytic viruses: a new class of immunotherapy drugs. Nat Rev Drug Discov 14:642 [Internet]. Nature Publishing Group. Available from: / pmc/articles/PMC7097180/. Cited 20 Aug 2022
15. Lin D, Feng X, Mai B, Li X, Wang F, Liu J et al (2021) Bacterial-based cancer therapy: an emerging toolbox for targeted drug/gene delivery. Biomaterials 277:121124 [Internet]. Available from: https://www.sciencedirect.com/science/article/pii/S0142961221004816
16. Khormi MA, Al-Maaqar SM, Al Johni AR, Al-Tayyar NA, Alhamad JA, Ghyathuddin AA et al (2025) Oncolytic bacteria: a revolutionary approach to cancer therapy. Open Life Sci 20:20251076
17. Kambouris M (2021) Bio-offense: black biology. In: Kambouris M (ed) Genomics in biosecurity, 1st edn. Elsevier Academic Press, London, pp 109–126
18. Kambouris ME, Georgoulas D (2021) Bio-offense: technical means, tactical approaches, operational orientations, and strategic concepts. In: Kambouris ME (ed) Genomics in biosecurity, 1st edn. Elsevier Academic Press, London, pp 127–142
19. Wittebole X, De Roock S, Opal SM (2014) A historical overview of bacteriophage therapy as an alternative to antibiotics for the treatment of bacterial pathogens. Virulence 5:226–235 [Internet]. Taylor and Francis Inc. Available from: https://pubmed.ncbi.nlm.nih.gov/23973944/. Cited 2 Mar 2021
20. Kambouris ME, Manoussopoulos Y, Kantzanou M, Velegraki A, Gaitanis G, Arabatzis M et al (2018) Rebooting bioresilience: a multi-OMICS approach to tackle global catastrophic biological risks and next-generation biothreats. Omics J Integr Biol 22:35–51
21. Acevedo-Rocha CG, Budisa N (2016) Xenomicrobiology: a roadmap for genetic code engineering. Microb Biotechnol 9:666–676 [Internet]. Wiley. Available from: https://pubmed.ncbi.nlm.nih.gov/27489097/. Cited 21 Sept 2020
22. Flandroy L, Poutahidis T, Berg G, Clarke G, Dao MC, Decaestecker E et al (2018) The impact of human activities and lifestyles on the interlinked microbiota and health of humans and of ecosystems. Sci Total Environ 627:1018–1038 [Internet]. Elsevier B.V. Available from: https://pubmed.ncbi.nlm.nih.gov/29426121/. Cited 21 Sept 2020
23. Diwo C, Budisa N (2018) Alternative biochemistries for alien life: basic concepts and requirements for the design of a robust biocontainment system in genetic isolation. Genes (Basel) 10 [Internet]. MDPI AG. Available from: http://www.ncbi.nlm.nih.gov/pubmed/30597824. Cited 21 Sept 2020
24. Budisa N, Kubyshkin V, Schmidt M (2020) Xenobiology: a journey towards parallel life forms. ChemBioChem 21:2228–2231 [Internet]. Wiley. Available from: https://doi.org/10.1002/cbic.202000141. Cited 11 Apr 2022
25. Schmidt M, Pei L, Budisa N (2018) Xenobiology: state-of-the-art, ethics, and philosophy of new-to-nature organisms. Adv Biochem Eng Biotechnol 162:301–316

26. Rummel JD (2001) Planetary exploration in the time of astrobiology: protecting against biological contamination. Proc Natl Acad Sci U S A 98:2128–2131
27. Hallsworth JE, Mancinelli RL, Conley CA, Dallas TD, Rinaldi T, Davila AF et al (2021) Astrobiology of life on earth. Environ Microbiol 23:3335–3344 [Internet]. Available from: https://pubmed.ncbi.nlm.nih.gov/33817931/. Cited 29 Nov 2022
28. Kambouris ME (2024) Beyond the biosphere. In: Biosecurity in the making: the threats, the aspects and the challenge of readiness. CRC Press, Boca Raton, pp 49–65
29. Kambouris ME (2021) Exploring the concepts: biosecurity, biodefence and biovigilance. In: Kambouris ME (ed) Genomics in biosecurity, 1st edn. Elsevier Academic Press, London, pp 3–14
30. Mix LJ (2015) Defending definitions of life. Astrobiology 15–19
31. Petkowski JJ, Bains W, Seager S (2020) On the potential of silicon as a building block for life. Life 10
32. Steele EJ (2020) Introduction—panspermia, 2020. Adv Genet 106:1 [Internet]. Elsevier. Available from: /pmc/articles/PMC7340401/. Cited 29 Nov 2022
33. Steele EJ, Gorczynski RM, Lindley RA, Tokoro G, Temple R, Wickramasinghe NC (2020) Origin of new emergent coronavirus and Candida fungal diseases—terrestrial or cosmic? Adv Genet 106:75–100 [Internet]. Elsevier. Available from: https://pubmed.ncbi.nlm.nih.gov/33081928/. Cited 29 Nov 2022
34. Feng W, Wan X, Zhang Y, Quensen J, Williams TA, Thompson M et al (2025) Diversification, niche adaptation, and evolution of a candidate phylum thriving in the deep critical zone. Proc Natl Acad Sci U S A 122:e2424463122
35. Wolfe-Simon F, Blum JS, Kulp TR, Gordon GW, Hoeft SE, Pett-Ridge J et al (2011) A bacterium that can grow by using arsenic instead of phosphorus. Science 332:1163–1166 [Internet]. Available from: https://pubmed.ncbi.nlm.nih.gov/21127214/. Cited 11 Apr 2022
36. Dadachova E, Casadevall A (2008) Ionizing radiation: how fungi cope, adapt, and exploit with the help of melanin. Curr Opin Microbiol 525–531
37. Passchier CW, Wassenaar TM, Groschopf N, Jantschke A, Mertz-Kraus R (2025) Subfossil fracture-related euendolithic micro-burrows in marble and limestone. Geomicrobiol J 42:390–405 [Internet]. Taylor & Francis. Available from: https://doi.org/10.1080/01490451.2025.2467417
38. Sengupta P, Muthamilselvi Sivabalan SK, Singh NK, Raman K, Venkateswaran K (2024) Genomic, functional, and metabolic enhancements in multidrug-resistant *Enterobacter bugandensis* facilitating its persistence and succession in the International Space Station. Microbiome 12:62
39. Pati NB, Jad SP, Schultze T, Mannala GK, Yao Y, Jaiswal S et al (2018) *Enterobacter bugandensis*: a novel enterobacterial species associated with severe clinical infection. Sci Rep 8:5392 [Internet]. Available from: https://doi.org/10.1038/s41598-018-23069-z
40. Yuan J, Zhang W, Dang L, Song Y, Yin Z, He Z et al (2025) *Niallia tiangongensis* sp. nov., isolated from the China Space Station. Int J Syst Evol Microbiol 75
41. Phillips C (1974) The planetary quarantine program: origins and achievements, 1956–1973. Scientific and Technical Information Office, National Aeronautics and Space Administration
42. Gupta RS (1998) Life's third domain (Archaea): an established fact or an endangered paradigm? Theor Popul Biol 54:91–104 [Internet]. Available from: http://www.ncbi.nlm.nih.gov/pubmed/9733652. Cited 5 July 2019
43. Woese CR, Kandler O, Wheelis ML (1990) Towards a natural system of organisms: proposal for the domains Archaea, Bacteria, and Eucarya. Proc Natl Acad Sci 87:4576–4579 [Internet]. Available from: http://www.ncbi.nlm.nih.gov/pubmed/2112744. Cited 10 June 2019
44. Pavia MJ, Garber AI, Avalle S, Macedo-Tafur F, Tello-Espinoza R, Cadillo-Quiroz H (2024) Functional insights of novel Bathyarchaeia reveal metabolic versatility in their role in peatlands of the Peruvian Amazon. Microbiol Spectr 12:e0038724
45. Poli A, Finore I, Romano I, Gioiello A, Lama L, Nicolaus B (2017) Microbial diversity in extreme marine habitats and their biomolecules. Microorganisms 5

46. Kosznik-Kwaśnicka K, Golec P, Jaroszewicz W, Lubomska D, Piechowicz L (2022) Into the unknown: microbial communities in caves, their role, and potential use. Microorganisms 10
47. Yong E (2014) Giant virus resurrected from 30,000-year-old ice. Nature [Internet]. Available from: https://doi.org/10.1038/nature.2014.14801
48. Rogers S, Starmer W, Castello J (2004) Recycling of pathogenic microbes through survival in ice. Med Hypotheses 63:733–737 [Internet]. Available from: https://pubmed.ncbi.nlm.nih.gov/15488645/. Cited 29 Nov 2022
49. Zhong ZP, Tian F, Roux S, Gazitúa MC, Solonenko NE, Li YF et al (2021) Glacier ice archives nearly 15,000-year-old microbes and phages. Microbiome 9:1–23 [Internet]. BioMed Central. Available from: https://doi.org/10.1186/s40168-021-01106-w. Cited 30 Nov 2022
50. Schultz J, dos Santos A, Patel N, Rosado AS (2023) Life on the edge: bioprospecting extremophiles for astrobiology. J Indian Inst Sci 103:721–737 [Internet]. Available from: https://doi.org/10.1007/s41745-023-00382-9
51. Currance P (2005) Decontamination procedures. In: Currance P (ed) Medical response to weapons of mass destruction. Elsevier Mosby, St. Louis, MI, pp 165–182
52. Ghafoor D, Khan Z, Khan A, Ualiyeva D, Zaman N (2021) Excessive use of disinfectants against COVID-19 posing a potential threat to living beings. Curr Res Toxicol 2:159–168
53. Wu Y, Yao M (2010) Inactivation of bacteria and fungus aerosols using microwave irradiation. J Aerosol Sci 41:682–693
54. Randall Harrell C, Djonov V, Fellabaum C, Volarevic V (2018) Risks of using sterilization by gamma radiation: the other side of the coin. Int J Med Sci 15:274–279
55. Alhmidi H, Cadnum JL, Piedrahita CT, John AR, Donskey CJ (2018) Evaluation of an automated ultraviolet-C light disinfection device and patient hand hygiene for reduction of pathogen transfer from interactive touchscreen computer kiosks. Am J Infect Control 46:464–467
56. Lopes RP, Mota MJ, Gomes AM, Delgadillo I, Saraiva JA (2018) Application of high pressure with homogenization, temperature, carbon dioxide, and cold plasma for the inactivation of bacterial spores: a review. Compr Rev Food Sci Food Saf 17:532–555
57. Basik AA, Sanglier J-J, Yeo CT, Sudesh K (2021) Microbial degradation of rubber: actinobacteria. Polymers 13
58. Velegraki A, Zerva L (2020) Identifying microbiota: genomic, mass-spectrometric, and serodiagnostic approaches. In: Kambouris M, Velegraki A (eds) Microbiomics: dimensions, applications, and translational implications of human and environmental microbiome research, 1st edn. Elsevier Academic Press, London, pp 77–94
59. David L, Brata AM, Mogosan C, Pop C, Czako Z, Muresan L et al (2021) Artificial intelligence and antibiotic discovery. Antibiotics 10
60. Beale CS, Runtuboy F, Sianipar AB, Beer AJE, Kadarusman, Erdmann MV et al (2025) Deep diving behaviour in oceanic manta rays and its potential function. Front Mar Sci 12 [Internet]. Available from: https://doi.org/10.3389/fmars.2025.1630451
61. Sharma P, Gero S, Payne R, Gruber DF, Rus D, Torralba A et al (2024) Contextual and combinatorial structure in sperm whale vocalisations. Nat Commun 15:3617 [Internet]. Available from: https://doi.org/10.1038/s41467-024-47221-8
62. Su Y, Shi S, Wang C, Wang Z, Li P, Zhang S et al (2024) Spider silk-inspired tough materials: multi-pathway synthesis, advanced processing, and functional applications. Nano Today 55:102188 [Internet]. Available from: https://www.sciencedirect.com/science/article/pii/S1748013224000434
63. Donati ER, Castro C, Urbieta MS (2016) Thermophilic microorganisms in biomining. World J Microbiol Biotechnol 32 [Internet]. Available from: https://pubmed.ncbi.nlm.nih.gov/27628339/. Cited 13 Dec 2021
64. Nogués-Bravo D, Simberloff D, Rahbek C, Sanders NJ (2016) Rewilding is the new Pandora's box in conservation. Curr Biol 26:R87–R91
65. Duckett D, Bjørkhaug H, Mur LA, Palmioli L (2022) New 'old' risks on the small farm: iconic species rewilding in Europe. Land Use Policy 118:106115

66. Sandler R (2017) De-extinction and conservation genetics in the Anthropocene. Hastings Cent Rep 47(Suppl 2):S43–S47
67. Díaz-Guerrero MÁ, Castillo-Juárez I, Zurabian R, Valdez A, Kota K, Hoshiko Y et al (2025) Reviving the past for a healthier future: ancient molecules and remedies as a solution to the antibiotic crisis. Future Microbiol 20:429–441 [Internet]. Taylor & Francis. Available from: https://doi.org/10.1080/17460913.2025.2476290
68. Schlebusch CM (2022) Genomics: testing the limits of de-extinction. Curr Biol 32:R324–R327 [Internet]. Elsevier. Available from: https://doi.org/10.1016/j.cub.2022.03.023
69. Tumpey T, Basler C, Aguilar P (2005) Characterization of the reconstructed 1918 Spanish influenza pandemic virus. Science 310:77–80
70. Osiro KO, Gil-Ley A, Fernandes FC, de Oliveira KBS, de la Fuente-Nunez C, Franco OL (2025) Paving the way for new antimicrobial peptides through molecular de-extinction. Microb Cell 12:1–8
71. Pappas G, Panagopoulou P, Akritidis N (2009) Reclassifying bioterrorism risk: are we preparing for the proper pathogens? J Infect Public Health 55–61
72. Kambouris ME, Velegraki A, Patrinos GP, Poulas K (2014) Reviewing the US select agents and toxins list with more efficient assessment of the potential applicability of microbes in bioterrorism: the fungal agent. Am J Bioterror Biosecur Biodef 1:3–5 [Internet]. Available from: www.austinpublishinggroup.com. Cited 15 Oct 2019
73. Casadevall A, Pirofski L (2004) The weapon potential of a microbe. Trends Microbiol 12:259–263 [Internet]. Available from: http://www.ncbi.nlm.nih.gov/pubmed/15165603. Cited 22 Mar 2019
74. Casadevall A, Pirofski L-A (2006) The weapon potential of human pathogenic fungi. Med Mycol 44:689–696 [Internet]. Available from: http://www.ncbi.nlm.nih.gov/pubmed/17127625. Cited 22 Mar 2019
75. Rubini R, Mayer C (2020) Addicting *Escherichia coli* to new-to-nature reactions. ACS Chem Biol 15:3093–3098 [Internet]. American Chemical Society. Available from: https://doi.org/10.1021/acschembio.0c00713. Cited 29 Nov 2022
76. Gumuskaya G, Srivastava P, Cooper BG, Lesser H, Semegran B, Garnier S et al (2024) Motile living biobots self-construct from adult human somatic progenitor seed cells. Adv Sci 11:2303575 [Internet]. Wiley. Available from: https://doi.org/10.1002/advs.202303575
77. Zhang H-K, Xu B-W, Jia Z-Y, Li B, Feng X-Q (2024) Inverse design of three-dimensional multicellular biobots with target functions. J Mech Phys Solids 187:105634 [Internet]. Available from: https://www.sciencedirect.com/science/article/pii/S0022509624001005
78. Özdemir V (2018) The dark side of the moon: the internet of things, industry 4.0, and the quantified planet. Omics J Integr Biol 22:637–641 [Internet]. Mary Ann Liebert, Inc.. Available from: https://doi.org/10.1089/omi.2018.0143. Cited 10 May 2019
79. Lokko Y, Heijde M, Schebesta K, Scholtès P, Van Montagu M, Giacca M (2018) Biotechnology and the bioeconomy—towards inclusive and sustainable industrial development. New Biotechnol 40:5–10
80. Sadler JC, Dennis JA, Johnson NW, Wallace S (2021) Interfacing non-enzymatic catalysis with living microorganisms. RSC Chem Biol 2:1073–1083 [Internet]. Royal Society of Chemistry. Available from: https://pubs.rsc.org/en/content/articlehtml/2021/cb/d1cb00072a. Cited 12 Apr 2022
81. Özdemir V, Hekim N (2018) Birth of industry 5.0: making sense of big data with artificial intelligence, "the internet of things" and next-generation technology policy. Omics J Integr Biol 22:65–76 [Internet]. Available from: https://doi.org/10.1089/omi.2017.0194. Cited 10 May 2019
82. McConnaughey M (2007) Life cycle of parasites. In: Enna SJ, Bylund DB (eds) xPharm: the comprehensive pharmacology reference. Elsevier, New York, pp 1–15. Available from: https://www.sciencedirect.com/science/article/pii/B9780080552323602326
83. Ding B (2010) Viroids: self-replicating, mobile, and fast-evolving noncoding regulatory RNAs. Wiley Interdiscip Rev RNA 1:362–375 [Internet]. Available from: https://pubmed.ncbi.nlm.nih.gov/21956936/. Cited 3 Nov 2020

84. Cavicchioli R, Ripple WJ, Timmis KN, Azam F, Bakken LR, Baylis M et al (2019) Scientists' warning to humanity: microorganisms and climate change. Nat Rev Microbiol 569–586 [Internet]. Nature Publishing Group. Available from: www.nature.com/nrmicro. Cited 16 Mar 2021
85. Lau SE, Teo WF, Teoh EY, Tan BC (2022) Microbiome engineering and plant biostimulants for sustainable crop improvement and mitigation of biotic and abiotic stresses. Discov Food 2:1–23 [Internet]. Springer. Available from: https://doi.org/10.1007/s44187-022-00009-5. Cited 26 Aug 2022
86. Liberton M, Page LE, O'Dell WB, O'Neill H, Mamontov E, Urban VS et al (2013) Organization and flexibility of cyanobacterial thylakoid membranes examined by neutron scattering. J Biol Chem 288:3632–3640
87. Picardo M, Filatova D, Nuñez O, Farré M (2019) Recent advances in the detection of natural toxins in freshwater environments. TrAC Trends Anal Chem 112:75–86
88. Whittaker RH (1969) New concepts of kingdoms of organisms. Science 163:150–160 [Internet]. American Association for the Advancement of Science. Available from: https://doi.org/10.1126/science.163.3863.150
89. Harada R, Nishimura Y, Nomura M, Yabuki A, Shiba K, Inaba K et al (2025) A cellular entity retaining only its replicative core: hidden archaeal lineage with an ultra-reduced genome. bioRxiv [Internet]. Available from: http://biorxiv.org/content/early/2025/05/02/2025.05.02.651781.abstract
90. Kambouris ME (2023) Global catastrophic biological risks in the post-COVID-19 world: time to act is now. OMICS 27:153–170 [Internet]. Available from: https://pubmed.ncbi.nlm.nih.gov/36946656/. Cited 24 July 2023
91. Raoult D, Forterre P (2008) Redefining viruses: lessons from Mimivirus. Nat Rev Microbiol 6:315–319
92. Forterre P (2006) The origin of viruses and their possible roles in major evolutionary transitions. Virus Res 117:5–16
93. Barthélémy R, Faure E, Goto T (2019) Serendipitous discovery in a marine invertebrate (Phylum Chaetognatha) of the longest giant viruses reported till date. Virol Curr Res 3
94. Williams AE, Sanchez-Vargas I, Reid WR, Lin J, Franz AWE, Olson KE (2020) The antiviral small-interfering RNA pathway induces zika virus resistance in transgenic *Aedes aegypti*. Viruses
95. Paludan SR, Pradeu T, Pichlmair A, Wray KB, Mikkelsen JG, Olagnier D et al (2024) Early host defense against virus infections. Cell Rep 43 [Internet]. Elsevier. Available from: https://doi.org/10.1016/j.celrep.2024.115070
96. Joyce GF, Szostak JW (2018) Protocells and RNA self-replication. Cold Spring Harb Perspect Biol 10
97. Koonin EV, Senkevich TG, Dolja VV (2006) The ancient virus world and evolution of cells. Biol Direct 1:29
98. Desnues C, Boyer M, Raoult D (2012) Sputnik, a virophage infecting the viral domain of life. In: Advances in virus research. Academic Press Inc., pp 63–89
99. Condit RC, Moussatche N, Traktman P (2006) In a nutshell: structure and assembly of the vaccinia virion. In: Advances in virus research, vol 66 [Internet]. Academic Press, pp 31–124. Available from: https://www.sciencedirect.com/science/article/pii/S0065352706660028
100. Liu L, Cooper T, Howley PM, Hayball JD (2014) From crescent to mature virion: vaccinia virus assembly and maturation. Viruses 6:3787–3808
101. deCarvalho T, Mascolo E, Caruso SM, López-Pérez J, Weston-Hafer K, Shaffer C et al (2023) Simultaneous entry as an adaptation to virulence in a novel satellite-helper system infecting *Streptomyces* species. ISME J 17:2381–2388
102. Hu CC, Hsu YH, Lin NS (2009) Satellite RNAs and satellite viruses of plants. Viruses 1:1325–1350 [Internet]. Available from: https://pubmed.ncbi.nlm.nih.gov/21994595/. Cited 20 Apr 2022
103. Koonin EV, Wolf YI (2012) Evolution of microbes and viruses: a paradigm shift in evolutionary biology? Front Cell Infect Microbiol 2:119

104. Koonin E (2010) The two empires and three domains of life in the postgenomic age. Nat Educ 3:27
105. Pearson H (2008) "Virophage" suggests viruses are alive. Nature 454:677 [Internet]. Nature Publishing Group. Available from: https://pubmed.ncbi.nlm.nih.gov/18685665/. Cited 22 Feb 2021
106. Dutta D, Ravichandiran V, Sukla S (2021) Virophages: association with human diseases and their predicted role as virus killers. Pathog Dis 79 [Internet]. Available from: https://pubmed.ncbi.nlm.nih.gov/34601577/. Cited 20 Aug 2022
107. Ogunrinola GA, Oyewale JO, Oshamika OO, Olasehinde GI (2020) The human microbiome and its impacts on health. Available from: https://doi.org/10.1155/2020/8045646
108. Geredew Kifelew L, Mitchell JG, Speck P (2019) Mini-review: efficacy of lytic bacteriophages on multispecies biofilms. Biofouling 1–10 [Internet]. Available from: https://doi.org/10.1080/08927014.2019.1613525. Cited 10 June 2019
109. Ursell LK, Metcalf JL, Parfrey LW, Knight R (2012) Defining the human microbiome. Available from: https://academic.oup.com/nutritionreviews/article/70/suppl_1/S38/1921538
110. Riccio P, Rossano R (2020) The human gut microbiota is neither an organ nor a commensal. FEBS Lett 594:3262–3271 [Internet]. Wiley. Available from: https://doi.org/10.1002/1873-3468.13946
111. Volland J-M, Gonzalez-Rizzo S, Gros O, Tyml T, Ivanova N, Schulz F et al (2022) A centimeter-long bacterium with DNA compartmentalized in membrane-bound organelles. bioRxiv [Internet]. Cold Spring Harbor Laboratory. Available from: https://doi.org/10.1101/2022.02.16.480423v1. Cited 22 Apr 2022
112. Nair A, Balasaravanan T, Jadhav S, Mohan V, Kumar C (2020) Harnessing the antibacterial activity of *Quercus infectoria* and *Phyllanthus emblica* against antibiotic-resistant *Salmonella typhi* and *Salmonella enteritidis* of poultry origin. Vet World 13:1388 [Internet]. Available from: /pmc/articles/PMC7429395/. Cited 30 Aug 2021
113. Purdel C, Ungurianu A, Adam-Dima I, Margină D (2023) Exploring the potential impact of probiotic use on drug metabolism and efficacy. Biomed Pharmacother 161:114468 [Internet]. Available from: https://www.sciencedirect.com/science/article/pii/S0753332223002561
114. Li XX, Yin J, Tang J, Li Y, Yang Q, Xiao Z et al (2018) Determining the balance between drug efficacy and safety by the network and biological system profile of its therapeutic target. Front Pharmacol 9:1245
115. Li Z, Lu G, Meng G (2019) Pathogenic fungal infection in the lung. Front Immunol 10:1524

Chapter 4
Synthetic Biology as a Major Biothreat Enhancer

Abstract In the context of the Meta-Biothreat, new breeds of pathogens may be used to derail surveillance and intervention. Such cases are the fully or partly engineered strains—Metapathogens, with six generations being identifiable. Synthetic biology, an engineering cognizant iteration of Applied Biology allows cognitive, methodological and operational optimization of such endeavors with cost- effective prototyping and production runs of very low to high outputs. Advanced options include genome pasting on minimal cellular chassis, recombining genomic elements of selected cells and virions, synthesizing microorganisms from commercially available (bio)chemical compounds, engineering xenomicroorganisms with non-canonical monomers in their macromolecules or entirely different macromolecules all the way to bionic microorganisms sporting the addition of mechanical parts in live cells, biocyborgs which combine normal and artificial cells and cyborgs and hybrid biobots with different mixtures of biological and technomechanical context and different qualities and capabilities. Such amenities pose extreme risks in terms of biosecurity and further the Meta-Biothreat risk manifold. Last but not least is the recurring concept of binary threats which refers to different formats, from independent triggers, biochemical or physical, that activate or reboot a live system (cellular organism, virus) to two-element cooperation as with helper viruses and, conceivably, to spatiotemporally coordinated epidemics.

Keywords Synthetic biology · Biohybrids · Biocyborgs · Metapathogens · Non-canonical aminoacids—ncAA · Xenonucleic acids—XNA · Hologenome · Overgenome–undergenome · Bacterial cell chassis

The Synthetic Biology is a concept somewhat misunderstood. It appeared as a counterpart of Synthetic Chemistry, meaning a growth of Biotechnology intended to "mass produce biological material amenities by purely biological processes and procedures, employed through engineering principles [1–3] for research purposes, manufacture of novel and/or high-value products and subsequently amplified to industrial scale" [4], whatever the term "industrial" might be taken to mean. The interpretation was rather wide. It is an engineering science [5] as it develops strategies for using composable

M. E. Kambouris, *Preparing for the Meta-Biothreat: Disruptive Biosecurity*, Advanced Sciences and Technologies for Security Applications,
https://doi.org/10.1007/978-3-032-25690-4_4

parts to build biological systems with predictable behaviors, not necessarily similar or equivalent to existing ones, but definitely of applicable nature [6]. Initially, there was no prerogative to create something entirely new, system or product. A procedural change, or rather optimization in production/processing, would qualify for Synthetic Biology achievement/product/endeavor.

The double inclusion of "biological" (end products/constructs/services and method of implementation) took Synthetic Biology apart from Biotechnology, which, although subject to different interpretations and definitions, would better be understood as "the use of biological resources, may these be material (organisms, or molecules) or immaterial (processes) to generate products or provide services" [7, 8]; a definition based on a very old and wide concept [9] and not incompatible with the one of the United Nations Convention on Biological Diversity (CBD) [10]. The enzymes contained in detergents for household laundry is a very nice example of the nature of Biotechnology at the interface of biological and the technomechanical/chemical realms.

The above definitions are handy but not free of problems: Bioreactors have been used for a long time to mass produce biomass or metabolites, and no-one ever called them Synthetic Biology, although it surely qualified as such. On the other hand, for eons Humanity used biological resources to achieve production or provision of services: the selective breeding, the use of animal power for hard labor or of microbiota for fermenting foodstuff or for bioremediation purposes [11], or the use of biofuel [6] are all activities that would qualify for inclusion under the term *Biotechnology*. The focus of both Synthetic Biology and Biotechnology to manipulation of nucleic acids and genetic/genomic engineering is not correct. It is just a dimension of both, the most visible and celebrated one, but other categories of both Synthetic Biology and Biotechnology do exist, such as protocells and bioinformatics respectively [2, 8, 12].

The core of the problem with Synthetic Biology as a Biothreat enhancer (a phrase used to substitute the much more restrictive "criminogenic potential", a favored term in Crime Studies and Law Enforcement [5] but failing to address its military and official political dimensions) comes from the quintessential quality of the discipline: the enhancement of producibility and diversification of bioentities [5, 13]. This might refer to different parameters: cost is one, but efficiency and environmental and technical footprint are two important other ones. Thus, Synthetic Biology allows production of bioagents, natural or engineered ones in massive scale in centralized facilities, but also in many, small, dispersed facilities with lower efficiency but also much lower footprint. The latter promotes eco-friendly production [14] as it makes away with the need to massively transport raw materials and products in and out of the processing compound from the sources of supply and to the users/customers respectively. Small quantities of raw materials are used locally to produce small outputs of goods to be distributed locally. It is a self-sufficiency enabler, excellent an investment for robustness of supply chains in case of crises and for market and economy robustness. Moreover, the dispersed parallel production enhances sustainability, as it alleviates the consumption of input resources (labor and all the support amenities, especially transportation, construction of accommodation and infrastructure and energy, water, connectivity and other utilities) in one site.

Synthetic Biology has many different guises; attempts to establish some order or even to properly enumerate them have not come to fruition. In this chapter the more relevant applications to the Meta-Biothreat will be presented and correlated to such malevolent use. These concern the manipulation of existing pathogens so as to acquire additional qualities and the development of new ones with such qualities integrated in their design [12]; both categories can be termed collectively "Metapathogens" (agents custom-engineered for pathogenicity), as they cannot be included among the known, existing pathogens, nor among the emerging but untampered ones which have diverse biological origins and niche/functions ("Neopathogens") [15, 16]. Such qualities are many and diverse [17]; still they mainly refer to defensive virulence factors, mainly but not exclusively resistance mechanisms to antibiotics but also to other antimicrobials like antiseptics, sterilizers, sanitizers, disinfectants and decontaminants [18, 19]; and/or additional toxicity and other offensive virulence factors as is the *Bacillus anthracis* toxin [20, 21]. A third category of factors refers to selectivity/host organism specificity and cell tropism, with the insertion of a four amino-acids sequence (PRRA) in the spike of SARS-COV-2 [22, 23] to be an example, and an indirect pointer to manipulation. This outcome is not due to nucleotide substitution but to an insertion (a four-aminoacids sequence requires a sequence of 12 nucleotides for its coding), of the exact suitable insert in the exact right position for maximum effectiveness, a really surgical alteration, very difficult to have occurred naturally. No party supporting the spontaneous nature of the SARS-CoV-2 creation has ever run the numbers for such possibilities.

4.1 Gain-of-Function

One of the first cases of Genetic Engineering was the introduction of the insulin code into plasmids so as to produce insulin in bacterial cells (*E. coli*) since 1979 and, after purification, make it available to diabetes patients [24]. It was immediately obvious that the method could be used to enhance a microbe with virulence factors, the paradigm being *Bacillus anthracis* and other cousin *Bacillus* species with the autologous virulence plasmids [25–27] and other select, rather mundane microbiota with xenologous genes. The latter case would be exemplified with the addition of the mouse Interleukin-4 gene in a poxvirus species selective for mice, an alteration that suppressed the immune response of the host and resulted in extremely high mortality of Mousepox [28]. This process, whence an organism is manipulated to acquire useful qualities by genetic engineering [20] was termed Gain-of-Function, to cloud the fact that it was a mutational technique for the development of more virile pathogens with combinations of characters highly unlikely to occur by natural recombination [29, 30], albeit not impossible. GoF was in essence development of biowarfare agents [31] and became legalized under the pretext of allowing research for understanding dangerous pathogens [2] and for developing countermeasures should such agents occur spontaneously or be developed by terrorists/non-state-players and some state players, defined as hostiles by some or other nefarious iteration of international community ethics and international laws. The justification could not quench the concern for destructive accidents, as the COVID-19 is alleged to have been [32], nor suppress the conclusion that there was a minimal distance from legitimized GoF endeavors to active, live bioagents' development [33, 34].

The GoF as explained above may refer to toxicity, but also to pathogenicity caused by lytic enzymes, which derail immune response or shape the tissue environment to the pathogen's needs. Such results are effected by their activity producing habitats, nutrients and also by physicochemical conditions as is pH, acidity and microatmosphere. But GoF may refer to defense mechanisms such as enzymes to neutralize or pumps to excrete antibiotics [20, 35], or spatial protection mechanisms: Such are sheath [36], glycocalyx [37], capsule or slime [38] and biofilm formation [39] to avert antibody attachment, complement rupture, phagocytosis and access of antimicrobials of any nature to the cell membrane. It may also refer to improved selectivity and power of adherence, the latter through slime, the former through receptors and other interfaces, such as the inserted PRRA aminoacid pattern of the SARS-CoV-2 spike protein [23].

4.2 Manufactured Microorganisms

The Synthetic Biology refers to many areas but mostly to fully engineered organisms, although parts of organisms and modified existing organisms are within its realm [12, 40]. For this reason its output is best described within the framework of the seven generations (0–6) of the newest bioagents' classification system [41] instead of the four-category previous one [42], despite their common traits [43]. Within the former classification system it applies to five or six of the seven generations: the First Generation is in doubt as it refers to selected, not modified agents and actually includes most of the western bioweapons until the 90s [44], the Japanese efforts up to the end of WW II [45] and also the first of the two generations of the Soviet BW assets, from 1927 to 1970 [46]. It also applies to 2 out of 4 categories recognized by the latter classification system, the Enhanced and the Advanced organisms, "Advanced" standing for wholly engineered ones.

The product of Synthetic Biology does not need to be innovative; it may be very common, but wholly synthesized ("bioengineered" [2]) from chemical compounds [47], which allows, in a Biosecurity context, to evade control regimes for hazardous microbial strains as enforced near the turn of the millennium, exemplified by the US Select Agents and Toxins list [2, 12, 48]. In the distant 2002 poliovirus cDNA was synthesized *de novo* from commercial biochemicals [49], and the whole genome of a minimal virus (φX174 phage) was synthesized from scratch a year later [47]. A much more complex and aggressive virus, in the Agrowarfare/Agroterrorism field of interest, although by no means exclusively so, the Horsepox virus was *de novo* synthesized in 2017 [50]. In this context, the SARS-CoV-2 may well have been modified, as mentioned above, or fully synthesized. Perhaps not from scratch, but through some more complicated pathway than the abovementioned GoF pipeline [32, 51], especially given (a) the blunt admittance of a high-profile publication for reverse engineering infectious SARS-CoV-2 cDNA during the pandemic "for research purposes" [52] and (b) the advent of Seamless Assembly since 2009 [53]. Seamless Assembly negates obvious signs of genetic manipulation, and by 2025 it evolved to the CRISPR-based PASTE method, applicable even to large DNA inserts [54]. The recent capability to use Adenoviruses, which allowed such insertions *in vivo* to non-dividing cells [55], is in essence a Fourth Generation genome attack agent [41] of tissue-specific targetability.

Regarding the bacteria, much more complex and evolved organisms and much more difficult to synthesize thereof [2], the first approach was to add genes and sequences—actually a GoF approach. The existence of plasmids did this an easier proposition [24], but insertion in the nucleoid (the main, circular bacterial DNA molecule which lacks the structure of the eukaryotic, proper chromosomes) was more challenging and more appealing [56], as it resulted in a more stable genetic alteration, a must for bioweapons that may be stored for long times, either in repetitive successive cultures or in frieze or other hibernation. A smarter solution was to paste a partial genome ("overgenome") [16] into the genetic space of a minimal genome ("undergenome") of some bacterial cell chassis [57]. This could be commercially [56] or selectively available for endeavors of organism and genomic engineering. The chassis would have most of the genes deleted but for the vital, housekeeping ones and thus when an overgenome would be pasted, a whole new genome ("hologenome") and a new organism would ensue [16]. The chassis could be genome-free and given a full new genome, either synthetically compiled [58], or assembled from sequences of different donor cells [59], or harvested intact from some other bacterial donor [60]. The technology of cell chassis for pasting existing, synthetic or hybrid genomes has evolved greatly, and the discipline reached the non-model chassis construction, i.e. fully reconfigurable chassis design [61], which is the counterpart of 3-D printing in mechanical engineering.

To produce a Eukaryote from scratch is infinitely more complex a task than with prokaryotes. Still, transplantation of nuclei between animal cells of different organisms has been an old experimental procedure and large swaths of the eukaryotic genome are either empty, although not junk [62, 63], or contain non-functional sequences, such as pseudogenes and deactivated proviruses [64, 65]. Thus there is available genomic space for pasting genetic information, if not by insertion, at least by recombination. By 2017 whole chromosomes for yeasts had been produced synthetically [66], an impossible task from a 90s aspect. Any editing or insertion in a wholly synthesized chromosome which will be delivered to a target nucleus subsequently, replacing the original one, is much more stable, predictable, accurately and precisely manipulated than interventions to existing chromosomes. With some numbers of such artificial chromosomes for a given genome, an overgenome/undergenome scheme [16] may become practical for Eukaryotes as well.

4.3 Xenobiology

Xenobiology has two pillars [67]: one is the Xenoorganisms of alien biochemistry and structure, in the context of Exo- and, especially, Astrobiology [68, 69]. The other is Xenobiota in Synthetic Biology [3], which refers to expanded, synthetic organisms fully engineered and only occasionally similar or equivalent to existing natural ones. Xenobiota are made up by analogues of the natural ingredients and procedures but in a more expansive nature: they are designed with genetic and potentially biochemical incompatibility with the biosphere [70, 71], in order to contain genetic pollution and forestall accidental interactions of detrimental nature, as when a xenobiote escapes containment for any reason, spontaneous or of human or cyber agency [72].

The xenobiota produced by Synthetic Biology demonstrate traits as required, are specifically designed for a given task with biocontainment and other security/safety features built into them from the beginning, as with synthetic auxotrophs [73]. In a biothreat iteration, there is no possibility of detection by usual molecular/spectrometric techniques as their respective qualities are (or may be) drastically altered compared to usual organisms, and unpredictably so. If biochemical compatibility is retained, by interfacing the xenogenetic code with the actual one, leading to conventional structural and functional macromolecules (proteins, lipids, sugars), there may be an opportunity for serological detection and classification. This would be the case whence an existing xenobiote, designed for biochemical compatibility with current biochemistry, for example to produce some high-value bioamenity with precision, stability and safety without the slavery of the natural producer cells (which may be proprietary), is aquired by malefactors, perhaps an inside operative, affiliated or not (as in industrial terrorism and crime) and modified for detrimental effects and nefarious purposes. Else, if the xenobiote is purpose-designed as a Bioweapon, even serological techniques may be unable to detect and classify it.

Even less probable is to read/sequence the xenogenetic code which may be made of non-canonical Nucleic Acids—ncNA, i.e. an existing, natural genetic code (among a number of such compiled by Nature) but with additional and/or modified nucleotides (non-canonical nucleotides), possibly but not necessarily combined with new proteomes made up wholly or partially by non-canonical aminoacids—ncAA [74]. Another level is the Xeno Nucleic Acids—XNA using chemically different residues and bonds between the chains and/or within the chain backbones. This may occur through different monomers (as with the threose-, hexose- or glycol-based nucleotides, instead of ribose/deoxyribose, which form TNA, HNA and GNA macromolecules respectively) that interact with Watson & Crick-based double helix bonds [72]. Or it might occur by different polymerization chemistry [2, 74–76]. Last, but not least, is the change of the current genetic code, retaining the same DNA and aminoacids by changing the tRNAs coupling of anticodons to aminoacids [69]. Irrespective of the exact nature of such codes, their structure and secrets lie with the designer and cannot be deduced at any stage by some screening/surveying method [17]. It is not only the ultimate biosecurity tool, but also the ultimate biothreat concept, unknown in shape and undetectable by existing methods. For example, even with identical base complementarity, the different backbones may render the lot of the DNA/RNA-based polymerases, nucleases and other relevant enzymes incompatible with the new substrate, as the 3-D electrochemistry of the backbone would be different, thus not allowing any detection/identification by NAAT and Restriction Enzyme Analysis-REA.

4.4 Complex Cases

This category refers to two different and diverse steps for producing an optimized Metapathogen. The second step always is of Synthetic Biology/Biotechnology, while the first might be also resulting from similar, Synthetic Biology origin, or not. It may be something completely different, like a natural event producing spontaneously a neopathogen or some non-pathogenic strain of interesting qualities. Exotic strains, naturally mutated strains, modified or wholly manufactured/artificial strains along with conventional ones, they all may be requisitioned from some legitimate source, as is an industrial facility, and further processed to turn Metapathogens.

Examples of natural events producing fast-track speciation are the recovery of radiotolerant microbiota found in the nuclear reactors of Chernobyl (to be accurate, rather radiosynthetic yeasts of Cryptococcal origin [77]) and the earthen microorganisms exposed for some time/generations to space conditions [78, 79]. Examples of putative neopathogens emerging as a result of *de novo* discovery are the exo-/astro-microorgannisms [68, 80, 81], and microbiomes of extreme and/or secluded environments (see Chap. 3).

Fast-track created receptor cells as per above—or raw materials in an engineering context—are by definition better attuned to tolerate ROS, pH, pressure, heat spike and thus survive immunoresponse, decon amenities or deployment hardships. Others, especially the ones coming from secluded environments and stellar bodies, are expected to be unchartered in terms of genome, chemical content (elementary/molecular) and shape/size/morphology. Thus they would be undetectable by targeted NAAT, probably by spectrometric methods and possibly by serological tests as well, and unidentifiable or unrecognizable by visual microscopy. Starting with these survivable/undetectable microorganisms, Synthetic Biology allows their manipulation to Meta- or Super-pathogens [16, 82], indicatively by inserting in extremophiles or in neopathogens some genetic blueprints for terminally effective virulence/effector factors such as lytic or other enzymes, bioregulators, and, especially, biotoxins.

4.5 Biocyborgs

The Cyborg, standing for CYBernetic ORganism, is understood to contain living tissues or even organs and mechanical parts and being controllable, the quintessence of Cybernetics. If one keeps the core idea of cybernetics, and replaces the mechanical parts with more biological parts (cellular bionics), or if the mechanical parts are simply deleted, there comes the Biocyborg. Ideally, some artificial cells would be integrated as they are more bound to accept control and behave predictably, since they are designed for such purpose from scratch. Thus, existing types of cells are used in an engineering set of tasks to operate as micromachines [83], prospectively but not exclusively within an organism, in close cooperation with the cells, tissues and organs of the latter. It is a case of chimerism with the chimeric cells being artificial, either custom-designed or modified—both being in the realm of Synthetic Biology. With these improvements an organism may be either cured from a number of conditions and syndromes, or endowed with extremely improved performance in various fields, including resilience and health [84]. It is a doorway to metahumans, with obvious consequences for military and security applications.

But simpler such chimerae, in the level of the microorganism, may be extremely appealing as bioagents. Such are plant pests for agroterrorism/agrowarfare [85], with the use of the fungus *Fusarium oxysporum* [86, 87] to eradicate Colombia's coca plants being focal although its implementation has been paused—allegedly at least, as in neighboring Bolivia legitimate such plantations were suffering severely by 2016 [86]. The above apply also to macroscopic meta-organisms, that would be weaponized by acquiring increased ferocity, endurance, power and awareness than current weaponized unmodified species, examples being dogs, falcons, dolphins and sharks for guarding, patrolling and search-and-destroy missions [88–91].

The use of the term "chimera" implies two distinct genomic pedigrees of cells within one organism, each with its own, separate genome; thus in essence this leaves out unicellular organisms. Though, with the biocyborgs, whence the artificial pedigree allows for more predictable applications and elaborate concepts, a more expanded notion may be used, to include three cases, or rather levels of fruitful co-existence.

The first level of such artificial-natural cell merge ("merge" being a better term than "hybridization") is the "population-level merge" whence the cells of two different pedigrees communicate almost freely, in cellular level, with one another across space, exchanging information and materials [83]. This is much like the usual chimerism and mosaicism, although the latter implies a genomic differentiation after the level of the zygote; it is a genomic alienation within the organism proper [92–94]. Another level, leading to a less conventional merge, is the "embedding", where normal and artificial cells are embedded inside one another. The encapsulated cells-symbionts perform organelle-like functions within their host. Both pedigrees may assume any of the two roles [83]. This is a rather unconventional use of the term "chimera" since the organism is unicellular and traditionally the genome is understood as the nuclear one of the host. But if the term "genome" is used to include all the genetic elements, including chromids and plasmids [95] in bacteria and occasionally fungi [96], and the exonuclear genomes, as in the eukaryotes that contain self-genomed organelles, namely plastids and mitochondria [97–99], then indeed the latter's genome does not stem from the same cellular progenitor as the nuclear constituent, and there is a notion of chimerism. The last level/case is a more structured, networked merge whence artificial and normal cells form higher structures, each of which contains one pedigree exclusively and they communicate as distinct entities that are physically linked to one another [83]. Such entities may be conglomerates, tissues or organs and the case is conceptually similar with the transplantation of solid organs [100].

4.6 Biohybrids

The biohybrids are a category that contains the cyborg but exceeds beyond it; the cyborg is a biohybrid, the biocyborg is not a biohybrid, the biohybrid just *may* be a cyborg. It refers to organisms containing both technomechanical and biological parts of any kind. Although it really reads uncomfortably, prosthetics, and especially motor or sensor—enabling prosthetics turn someone into a biohybrid but not into a cyborg. Even silicone implants and screwed teeth surrogates qualify for this characterization, not to mention pacemakers and other control devices. The difference between cyborgs and biohybrids may be in the nature of used parts, but it is suggested herein that it has much to do with control. Cyborgs are cybernetic organisms, meaning they are made and meant to be controlled by another decision-maker. Biohybrids are not necessarily so. The thin but clear line may be the focus point of neurohacking [5]. Thus, the remote control of insect flights [101] is neurohacking and the insects modified with prosthetics, neurohacked to send data from their olfactory sensors to their controller and being actively controlled by them to be sent to sniff explosives are an exemplary case of neurohacked biohybrid [102].

The biobots are controllable organisms of a single cell pedigree/genome, as opposed to the biocyborgs. If the biobot structurally and functionally combines with technomechanical substrate, then comes the Hybrid biobot, which is a cybernetic organism of mixed biological and technomechnical construction, but has to include functional biological parts essential for its function, whereas in the cyborg the bioparts may be used for other purposes, including aesthetics. Thus, biocompatible technomechanical skeletons that support and morph in 3-D aggregations of cells constitute biobots [103, 104] and could be used to effectively disseminate effectors to degrade (or support or reconstitute) live, dead or inanimate matter rather than the physical shape and bodyform of the original cell source organism, which is highly unoptimized for peculiar tasks. Biological evolution opted for survival, not for the efficiency of an organism as an instrument of therapy, or war.

The Biote-bot, on the other hand, is a fully integrated but conditional merge of microrobots and actual cells, the latter being natural, enhanced or fully engineered, to perform an assigned mission either under AI guidance, or through remote control—a concept most adventurous but possibly fictious up to now [105]. The technology is meant for beneficial delicate operations such as cell-scale delivery and repairs or replacement/substitution, but has great criminogenic and martial potential. It is perfect for clandestine operations with very predictable results of seemingly natural causality.

A special case is the sensor cells that may use technomechanical or bioengineered solutions to sense some moiety in the environment and then produce a readable signal.

In theory this is embedded in all bioorganisms: in the presence of nutrients they grow, higher organisms when met with favorable environment show some kind of approval, comfort and ease. Something more elaborate, such as some visual/color signal, but also other approaches, would turn such organisms to detectors of completely alien substances—in relation to their niche and habitat [96, 106].

4.7 Binary Threats

The idea of binary effectors is, once more, not a new one. Since its inception it allowed modularity, but most importantly, it enhanced safety and security during the multiple steps to be taken before the initiation of active use/deployment of a weapon, especially the steps required for the immediate preparation of the said use. The binary nature makes more difficult to apperceive or at least to correctly assess an assault. In one example of this attack method, one platform acquired the target to guide the weapon fired/released by another platform. The modern paradigm has been the cooperative attacks that allowed aircraft strikes with elaborate approach and escape patterns and thus high survival rates. Supposedly it was initiated by the USA armed forces: first, the "buddy lasing" by USAF aircraft carrying AVQ-23 Pave Spike or AVQ-10 Pave Knife electrooptical targeting pods, which included laser illuminators, so as to "lase" surface targets for attacks with laser-guided weapons dropped by other aircraft. Such weapons were the GBU-2/5 Pavestorm cluster bombs and the GBU-1/10/11/12 Paveway unitary bombs. The idea continued with the M-712 Copperhead 155 mm artillery shells guided by laser illuminators used by infantry and Special Forces, a concept taken by the Soviets with the Krasnopol and Kitolov laser guided artillery and mortar shells. More applications ensued in the 80s [107, 108].

The term "binary weapon" though was coined for the USAF's air-dropped BLU-80 Bigeye chemical warfare 500-lbs bombs of the 80s. It is unclear if they ever reached Operational Capability, but in theory they were to release the VX nerve agent that was produced within the bomb during its flight, when two separate containers, each carrying one benign precursor, were ruptured and their contents mixed to react and produce the VX [109, 110].

In the context of bioagents, a binary system implies that once delivered, the effector agent needs another procedural step, an enabler or trigger, so as to become active [107]. The activation may be of chemical [40] or physical causality [111, 112], digital in nature, as is the transition to vegetative metabolism from a previous stasis/dormancy/latency for organisms, or to active from some other non-functional, precursor status for molecules. It may also be quantitative/qualitative, as in the change of the life cycle phase of a microorganism by a bioregulator, for example in order to initiate sporulation, or toxin production, or propagation.

Another possibility is to enhance the agent *in situ*, actually to stimulate its propagation so as to increase the lethal doses manifold despite initially dispersing or planting mush smaller quantities, the latter being easier to handle and hide. For organisms, induction effects may be caused by chemical stimuli [40] or by physical ones, conceivably sound, color/light or other irradiation, and definitely temperature manipulation [113] or exposure to electricity or magnetism [111, 114]. For biochemical/molecular agents this is effected by an enzyme or a co-factor that would change the 3-D conformation of the effector molecule to higher activity.

And then a benign agent, either dispersed or already in place as in local or appendage microbiomes, may be transformed into malign form *in situ*: this could happen conceivably by DNA enhancement through Lateral Gene Transfer, meaning transformation, transposition, conjugation or viral transduction, in both Bacteria and Eucarya [115, 116]. There are unconfirmed reports that the Soviets had binary systems of bacteria and plasmids with virulent factors and mixed them during the flight of the warhead, supposedly in an Intercontinental Ballistic Missile (ICBM), so by the time of the dispersion a much more virulent bioagent would have been created [107, 117].

Last but not least is the succession of two apparently independent outbreaks/events as detailed in Chap. 5. It is not a binary event in the true sense, in that both events may occur independently and cause detrimental results each one by itself. But, especially in the context of the Meta-Biothreat, they are interdependent, whatever their spatiotemporal distance may be in absolute metrics. The interdependence is judged by their synergistic effect, i.e. how the combined effect is more severe than the sum of the effects of the two events. They may be arranged so as to deplete human and material assets of Medical CounterMeasures (MCM), or so as the follow-on event capitalizes on and exploits a given direction and commitment of the immunity, surveillance and intervention effected by the need to respond to the initial event. The Monkeypox is considered by itself a possible bioassault risk [118] but the occurrence of a limited such outbreak, with minimal destructiveness, in the wake of the COVID-19, before the latter had died out as a pandemic [119, 120], and thus partially overlapping spatiotemporally with it, amplified the former's impact due to surging/sequential effects [121]. In this case, both events were viral ones, allegedly of animal origin and of spontaneous causality. In the future, two overlapping or surging events of entirely different character may prove more of a challenge, and even more so if they are not just two, but any number combined in any spatiotemporal pattern; a Multiplicity of perpetrated Bioevents [41, 43].[1]

References

1. Decoene T, De Paepe B, Maertens J, Coussement P, Peters G, De Maeseneire SL et al (2018) Standardization in synthetic biology: an engineering discipline coming of age. Crit Rev Biotechnol 647–656

[1] For a more thorough discussion of the binary threat the interested reader is prompted to take into consideration the Book Hybrid warfare 2.2, Chap. 6 Sect. 4 pp. 111–115, as the true essence of such practice lies with the Hybrid Warfare concept rather than with the Meta-Biothreat. In here the excerpt is included to properly round off the issue of Synthetic Biology aspects of the Meta-Biothreat.

2. Gómez-Tatay L, Hernández-Andreu JM (2019) Biosafety and biosecurity in synthetic biology: a review. Crit Rev Environ Sci Technol 49:1587–1621 [Internet]. Taylor and Francis Inc. Available from: https://doi.org/10.1080/10643389.2019.1579628. Cited 18 Sept 2020
3. Benner SA, Sismour AM (2005) Synthetic biology. Nat Rev Genet 6:533–543 [Internet]. Available from: https://doi.org/10.1038/nrg1637
4. Shi T, Han P, You C, Zhang Y-HPJ (2018) An in vitro synthetic biology platform for emerging industrial biomanufacturing: bottom-up pathway design. Synth Syst Biotechnol 3:186–195 [Internet]. Available from: https://www.sciencedirect.com/science/article/pii/S2405805X18300255
5. Elgabry M, Nesbeth D, Johnson SD (2020) A systematic review of the criminogenic potential of synthetic biology and routes to future crime prevention. Front Bioeng Biotechnol 8:571672
6. Carlson R (2007) Laying the foundations for a bio-economy. Syst Synth Biol 1:109–117 [Internet]. Springer. Available from: http://www.ncbi.nlm.nih.gov/pubmed/19003445. Cited 10 May 2019
7. Abbate E, Andrion J, Apel A, Biggs M, Chaves J, Cheung K et al (2023) Optimizing the strain engineering process for industrial-scale production of bio-based molecules. J Ind Microbiol Biotechnol 50
8. Bentahar S, Abada R, Ykhlef N (2023) Biotechnology: definitions, types and main applications. YMER 22:563–575
9. Ereky K (1919) Biotechnologie der Fleisch-, Fett-, und Milcherzeugung im landwirtschaftlichen Grossbetriebe: für naturwissenschaftlich gebildete Landwirte verfasst. P. Parey, Berlin
10. Okafor N, Okeke B (2017) Modern industrial microbiology and biotechnology. CRC Press, Boca Raton
11. Janssen DB, Stucki G (2020) Perspectives of genetically engineered microbes for groundwater bioremediation. Environ Sci Process Impacts 22:487–499 [Internet]. Royal Society of Chemistry. Available from: http://www.pubchem.ncbi.nlm.nih.gov. Cited 16 Mar 2021
12. Lentzos F (2012) Synthetic biology, security and governance. Biosocieties 7:339–351 [Internet]. Available from: http://www.ncbi.nlm.nih.gov/pubmed/32226471. Cited 21 Sept 2020
13. Choi KR, Jang WD, Yang D, Cho JS, Park D, Lee SY (2019) Systems metabolic engineering strategies: integrating systems and synthetic biology with metabolic engineering. Trends Biotechnol 817–837
14. Ezeako EC, Nwiloh BI, Odo MC, Ozougwu VE (2025) Harnessing synthetic biology for sustainable industrial innovation: advances, challenges, and future direction. Biochem Eng J 221:109777 [Internet]. Available from: https://www.sciencedirect.com/science/article/pii/S1369703X25001512
15. Kambouris M (2021) The concept of humanome and the microbiomic dimension. In: Kambouris M (ed) Genomics in biosecurity, 1st edn. Elsevier Academic Press, London, pp 15–28
16. Kambouris ME, Manoussopoulos Y, Kantzanou M, Velegraki A, Gaitanis G, Arabatzis M et al (2018) Rebooting bioresilience: a multi-OMICS approach to tackle global catastrophic biological risks and next-generation biothreats. Omics J Integr Biol 22:35–51
17. Kambouris M (2021) Bio-offense: black biology. In: Kambouris M (ed) Genomics in biosecurity, 1st edn. Elsevier Academic Press, London, pp 109–126
18. McCarlie S, Boucher CE, Bragg RR (2020) Molecular basis of bacterial disinfectant resistance. Drug Resist Updat 48
19. Vijayakumar R, Sandle T (2019) A review on biocide reduced susceptibility due to plasmid-borne antiseptic-resistant genes—special notes on pharmaceutical environmental isolates. J Appl Microbiol 126:1011–1022
20. Gargis AS, Cherney B, Conley AB, Mclaughlin HP, Sue D (2019) Rapid detection of genetic engineering, structural variation, and antimicrobial resistance markers in bacterial biothreat pathogens by nanopore sequencing. Nat Sci Rep 9:13501 [Internet]. Available from: https://doi.org/10.1038/s41598-019-49700-1. Cited 1 July 2023

21. Pena-Gonzalez A, Rodriguez-R LM, Marston CK, Gee JE, Gulvik CA, Kolton CB et al (2018) Genomic characterization and copy number variation of *Bacillus anthracis* plasmids pXO1 and pXO2 in a historical collection of 412 strains. mSystems 3
22. Argyrou M, Pitsillou E, Hung A, El-Osta A, Karagiannis TC (2025) Insights into the pathogenic mechanisms associated with the SARS-CoV-2 spike protein. J Struct Biol 217:108229 [Internet]. Available from: https://www.sciencedirect.com/science/article/pii/S1047847725000644
23. Liu S, Selvaraj P, Lien CZ, Nunez IA, Wu WW, Chou C-K et al (2021) The PRRA insert at the S1/S2 site modulates cellular tropism of SARS-CoV-2 and ACE2 usage by the closely related bat RaTG13. J Virol 95
24. Baeshen NA, Baeshen MN, Sheikh A, Bora RS, Ahmed MMM, Ramadan HAI et al (2014) Cell factories for insulin production. Microb Cell Fact 13 [Internet]. BioMed Central. Available from: /pmc/articles/PMC4203937/. Cited 30 Nov 2022
25. Sahl JW, Pearson T, Okinaka R, Schupp JM, Gillece JD, Heaton H et al (2016) A *Bacillus anthracis* genome sequence from the Sverdlovsk 1979 autopsy specimens. MBio 7 [Internet]. American Society for Microbiology. Available from: https://pubmed.ncbi.nlm.nih.gov/27677796/. Cited 3 Nov 2020
26. Furuta Y, Cheng C, Zorigt T, Paudel A, Izumi S, Tsujinouchi M et al (2021) Direct regulons of AtxA, the master virulence regulator of *Bacillus anthracis*. mSystems 6:e0029121
27. Taylor-Joyce G, Manoharan S, Brooker TA, Hernández-Rodríguez CS, Baillie L, Oyston PCF et al (2023) The influence of extrachromosomal elements in the anthrax "cross-over" strain *Bacillus cereus* G9241. Front Microbiol 14:1113642
28. Jackson RJ, Ramsay AJ, Christensen CD, Beaton S, Hall DF, Ramshaw IA (2001) Expression of mouse interleukin-4 by a recombinant ectromelia virus suppresses cytolytic lymphocyte responses and overcomes genetic resistance to mousepox. J Virol 75:1205–1210 [Internet]. American Society for Microbiology. Available from: https://pubmed.ncbi.nlm.nih.gov/11152493/. Cited 21 Sept 2020
29. Pérez-Losada M, Arenas M, Galán JC, Palero F, González-Candelas F (2015) Recombination in viruses: mechanisms, methods of study, and evolutionary consequences. Infect Genet Evol 30:296 [Internet]. Elsevier. Available from: /pmc/articles/PMC7106159/. Cited 22 Apr 2022
30. Hanage WP (2016) Not so simple after all: bacteria, their population genetics, and recombination. Cold Spring Harb Perspect Biol 8 [Internet]. Cold Spring Harbor Laboratory Press. Available from: /pmc/articles/PMC4930924/. Cited 16 Mar 2021
31. Chen D-Y, Kenney D, Chin CV, Tavares AH, Khan N, Conway HL et al (2022) Role of spike in the pathogenic and antigenic behavior of SARS-CoV-2 BA.1 Omicron. bioRxiv [Internet]. Cold Spring Harbor Laboratory. Available from: https://doi.org/10.1101/2022.10.13.512134v1. Cited 29 Nov 2022
32. Holmes EC, Goldstein SA, Rasmussen AL, Robertson DL, Crits-Christoph A, Wertheim JO et al (2021) The origins of SARS-CoV-2: a critical review. Cell 184:4848–4856 [Internet]. Elsevier. Available from: /pmc/articles/PMC8373617/. Cited 5 Aug 2022
33. Kaiser J (2014) Scientists call for limit on creating dangerous pathogens. Science [Internet]. Available from: https://www.science.org/content/article/scientists-call-limit-creating-dangerous-pathogens
34. Risk and benefit analysis of gain of function research (2015). Draft final report. Takoma Park, MD
35. Howden BP, McEvoy CRE, Allen DL, Chua K, Gao W, Harrison PF et al (2011) Evolution of multidrug resistance during *Staphylococcus aureus* infection involves mutation of the essential two component regulator WalKR. PLoS Pathog 7 [Internet]. Available from: https://pubmed.ncbi.nlm.nih.gov/22102812/. Cited 19 Apr 2022
36. Springstein B, Weissenbach J, Koch R, Stücker F, Stucken K (2020) The role of the cytoskeletal proteins MreB and FtsZ in multicellular cyanobacteria. FEBS Open Bio 10:2510–2531 [Internet]. Available from: https://pubmed.ncbi.nlm.nih.gov/33112491/. Cited 20 Oct 2021
37. Wargo MJ, Hogan DA (2006) Fungal-bacterial interactions: a mixed bag of mingling microbes. Curr Opin Microbiol 9:359–364

38. Makvandi P, Song H, Yiu CKY, Sartorius R, Zare EN, Rabiee N et al (2023) Bioengineered materials with selective antimicrobial toxicity in biomedicine. Mil Med Res 10 [Internet]. Available from: https://pubmed.ncbi.nlm.nih.gov/36829246/. Cited 29 July 2023
39. Adam B, Baillie GS, Douglas LJ (2002) Mixed species biofilms of *Candida albicans* and *Staphylococcus epidermidis*. J Med Microbiol 51:344–349 [Internet]. Available from: https://doi.org/10.1099/0022-1317-51-4-344. Cited 18 Feb 2019
40. Ford TJ, Silver PA (2015) Synthetic biology expands chemical control of microorganisms. Curr Opin Chem Biol 28:20–28
41. Kambouris ME (2023) Global catastrophic biological risks in the post-COVID-19 world: time to act is now. OMICS 27:153–170 [Internet]. Available from: https://pubmed.ncbi.nlm.nih.gov/36946656/. Cited 24 July 2023
42. Colf LA (2016) Preparing for nontraditional biothreats. Health Secur 14:7–12 [Internet]. Available from: https://pubmed.ncbi.nlm.nih.gov/26863312/. Cited 22 Apr 2022
43. Kambouris ME (2024) The GCBR after COVID-19. In: Hybrid warfare 2.2: where biothreats meet irregular operations and cyber warriors in the 21st century, 1st edn. Springer, Cham, pp 143–186
44. Anderson PD, Bokor G (2012) Bioterrorism: pathogens as weapons. J Pharm Pract 25:521–529 [Internet]. Sage. Available from: https://pubmed.ncbi.nlm.nih.gov/23011963/. Cited 22 Feb 2021
45. Keiichi T (2005) Unit 731 and the Japanese Imperial Army's biological warfare program. Asia-Pac J 3:2194 [Internet]. Available from: https://apjjf.org/-Tsuneishi-Keiichi/2194/article.html. Cited 3 Nov 2020
46. Zilinskas R (2016) The Soviet biological weapons program and its legacy in today's Russia. Report No.: 11. Washington, DC
47. Smith HO, Hutchison CA, Pfannkoch C, Venter JC (2003) Generating a synthetic genome by whole genome assembly: φX174 bacteriophage from synthetic oligonucleotides. Proc Natl Acad Sci U S A 100:15440–15445 [Internet]. Available from: https://pubmed.ncbi.nlm.nih.gov/14657399/. Cited 21 Sept 2020
48. Kambouris ME, Velegraki A, Patrinos GP, Poulas K (2014) Reviewing the US select agents and toxins list with more efficient assessment of the potential applicability of microbes in bioterrorism: the fungal agent. Am J Bioterror Biosecur Biodef 1:3–5 [Internet]. Available from: www.austinpublishinggroup.com. Cited 15 Oct 2019
49. Cello J, Paul AV, Wimmer E (2002) Chemical synthesis of poliovirus cDNA: generation of infectious virus in the absence of natural template. Science 297:1016–1018 [Internet]. Available from: https://pubmed.ncbi.nlm.nih.gov/12114528/. Cited 2 Nov 2020
50. Koblentz GD (2017) The de novo synthesis of horsepox virus: implications for biosecurity and recommendations for preventing the reemergence of smallpox. Health Secur 15:620–628
51. Segreto R, Deigin Y (2021) The genetic structure of SARS-CoV-2 does not rule out a laboratory origin: SARS-COV-2 chimeric structure and furin cleavage site might be the result of genetic manipulation. Bioessays 43 [Internet]. Available from: https://pubmed.ncbi.nlm.nih.gov/33200842/. Cited 7 Sept 2023
52. Xie X, Lokugamage KG, Zhang X, Vu MN, Muruato AE, Menachery VD et al (2021) Engineering SARS-CoV-2 using a reverse genetic system. Nat Protoc 16:1761–1784 [Internet]. Available from: https://doi.org/10.1038/s41596-021-00491-8
53. Gibson DG, Young L, Chuang R-Y, Venter JC, Hutchison CA 3rd, Smith HO (2009) Enzymatic assembly of DNA molecules up to several hundred kilobases. Nat Methods 6:343–345
54. Gwon LW, Badon IW, Lee Y, Kim H-J, Lee SH (2025) Advances in large-scale DNA engineering with the CRISPR system. Exp Mol Med 57:1902–1912 [Internet]. Available from: https://doi.org/10.1038/s12276-025-01530-0
55. Estes BJG, Gandhi N, Von Stetina JR, Paudel D, Nan AX, Amin P et al (2025) Development of circular AAV cargos for targeted seamless insertion with large serine integrases. Mol Ther Methods Clin Dev 33:101490
56. Bayer CN, Rennig M, Ehrmann AK, Nørholm MHH (2021) A standardized genome architecture for bacterial synthetic biology (SEGA). Nat Commun 12:5876 [Internet]. Available from: https://doi.org/10.1038/s41467-021-26155-5

57. Hutchison CA, Chuang RY, Noskov VN, Assad-Garcia N, Deerinck TJ, Ellisman MH et al (2016) Design and synthesis of a minimal bacterial genome. Science 351
58. Gibson DG, Glass JI, Lartigue C, Noskov VN, Chuang RY, Algire MA et al (2010) Creation of a bacterial cell controlled by a chemically synthesized genome. Science 329:52–56
59. Zhu M-C, Cui Y-Z, Wang J-Y, Xu H, Li B-Z, Yuan Y-J (2023) Cross-species microbial genome transfer: a review. Front Bioeng Biotechnol 11 [Internet]. Available from: https://doi.org/10.3389/fbioe.2023.1183354
60. Lartigue C, Glass JI, Alperovich N, Pieper R, Parmar PP, Hutchison CA 3rd et al (2007) Genome transplantation in bacteria: changing one species to another. Science 317:632–638
61. Hwang S, Joung C, Kim W, Palsson B, Cho B-K (2023) Recent advances in non-model bacterial chassis construction. Curr Opin Syst Biol 36:100471 [Internet]. Available from: https://www.sciencedirect.com/science/article/pii/S2452310023000288
62. Story MD, Durante M (2018) Radiogenomics. Med Phys 45:e1111–e1122 [Internet]. Available from: https://pubmed.ncbi.nlm.nih.gov/30421807/. Cited 6 Dec 2021
63. Cech TR, Steitz JA (2014) The noncoding RNA revolution-trashing old rules to forge new ones. Cell 157:77–94 [Internet]. Available from: https://doi.org/10.1016/j.cell.2014.03.008. Cited 11 Apr 2024
64. Velegraki A, Kambouris ME (2011) The adaptive genome: a hypothesis and its possible impact on diagnostic methodology. Arch Hell Med 28:317–322
65. Zhang Z, Gerstein M (2003) Patterns of nucleotide substitution, insertion and deletion in the human genome inferred from pseudogenes. Nucleic Acids Res 31:5338–5348 [Internet]. Oxford University Press. Available from: http://www.ncbi.nlm.nih.gov/pubmed/12954770. Cited 28 Feb 2019
66. Ma Y, Zhang Z, Jia B, Yuan Y (2024) Automated high-throughput DNA synthesis and assembly. Heliyon 10:e26967
67. Schmidt M, Pei L, Budisa N (2018) Xenobiology: state-of-the-art, ethics, and philosophy of new-to-nature organisms. Adv Biochem Eng Biotechnol 162:301–316
68. Kambouris ME (2024) Beyond the biosphere. In: Biosecurity in the making: the threats, the aspects and the challenge of readiness. CRC Press, Boca Raton, pp 49–65
69. Diwo C, Budisa N (2018) Alternative biochemistries for alien life: basic concepts and requirements for the design of a robust biocontainment system in genetic isolation. Genes 10 [Internet]. MDPI AG. Available from: http://www.ncbi.nlm.nih.gov/pubmed/30597824. Cited 21 Sept 2020
70. Budisa N, Kubyshkin V, Schmidt M (2020) Xenobiology: a journey towards parallel life forms. ChemBioChem 21:2228–2231 [Internet]. Wiley. Available from: https://doi.org/10.1002/cbic.202000141. Cited 11 Apr 2022
71. Völler JS, Budisa N (2017) Coupling genetic code expansion and metabolic engineering for synthetic cells. Curr Opin Biotechnol 1–7
72. Schmidt M (2010) Xenobiology: a new form of life as the ultimate biosafety tool. BioEssays 32:322–331 [Internet]. Available from: https://pubmed.ncbi.nlm.nih.gov/20217844/. Cited 21 Sept 2020
73. Rubini R, Mayer C (2020) Addicting *Escherichia coli* to new-to-nature reactions. ACS Chem Biol 15:3093–3098 [Internet]. American Chemical Society. Available from: https://doi.org/10.1021/acschembio.0c00713. Cited 29 Nov 2022
74. Acevedo-Rocha CG, Budisa N (2016) Xenomicrobiology: a roadmap for genetic code engineering. Microb Biotechnol 9:666–676 [Internet]. Wiley. Available from: https://pubmed.ncbi.nlm.nih.gov/27489097/. Cited 21 Sept 2020
75. Schmidt M, Pei L (2011) Synthetic toxicology: where engineering meets biology and toxicology. Toxicol Sci 120:S204–S224 [Internet]. Oxford Academic. Available from: https://academic.oup.com/toxsci/article/120/suppl_1/S204/1617750. Cited 1 Dec 2022
76. Kubyshkin V, Budisa N (2017) Synthetic alienation of microbial organisms by using genetic code engineering: why and how? Biotechnol J 12
77. Dadachova E, Casadevall A (2008) Ionizing radiation: how fungi cope, adapt, and exploit with the help of melanin. Curr Opin Microbiol 525–531

78. Yuan J, Zhang W, Dang L, Song Y, Yin Z, He Z et al (2025) *Niallia tiangongensis* sp. nov., isolated from the China Space Station. Int J Syst Evol Microbiol 75
79. Sengupta P, Muthamilselvi Sivabalan SK, Singh NK, Raman K, Venkateswaran K (2024) Genomic, functional, and metabolic enhancements in multidrug-resistant *Enterobacter bugandensis* facilitating its persistence and succession in the International Space Station. Microbiome 12:62
80. Hallsworth JE, Mancinelli RL, Conley CA, Dallas TD, Rinaldi T, Davila AF et al (2021) Astrobiology of life on earth. Environ Microbiol 23:3335–3344 [Internet]. Available from: https://pubmed.ncbi.nlm.nih.gov/33817931/. Cited 29 Nov 2022
81. Kambouris ME (2021) Exploring the concepts: biosecurity, biodefence and biovigilance. In: Kambouris ME (ed) Genomics in biosecurity, 1st edn. Elsevier Academic Press, London, pp 3–14
82. Kambouris ME, Gaitanis G, Manoussopoulos Y, Arabatzis M, Kantzanou M, Kostis GD et al (2018) Humanome versus microbiome: games of dominance and pan-biosurveillance in the omics universe. Omics J Integr Biol 22:528–538
83. Elani Y (2021) Interfacing living and synthetic cells as an emerging frontier in synthetic biology. Angew Chemie Int Ed 60:5602–5611 [Internet]. Wiley. Available from: https://doi.org/10.1002/anie.202006941. Cited 6 Oct 2022
84. Pio-Lopez L (2021) The rise of the biocyborg: synthetic biology, artificial chimerism and human enhancement. New Genet Soc 40:599–619 [Internet]. Available from: https://www.tandfonline.com/action/journalInformation?journalCode=cngs20
85. Olsen A (2003) U.S. moves towards biological warfare in Colombia. PANNA [Internet]. Available from: https://www.panna.org/legacy/panups/panup_20030124.dv.html. Cited 23 Aug 2022
86. Pearson Z (2016) "Coca got us here and now it's our weakness:" *Fusarium oxysporum* and the political ecology of a drug war policy alternative in Bolivia. Int J Drug Policy 33:88–95
87. de Vries G (2000) *Fusarium* considered to kill coca plants. Trends Plant Sci 5:417 [Internet]. Elsevier. Available from: https://doi.org/10.1016/S1360-1385(00)01774-X
88. Nikolić SS (2017) An innovative response to commercial UAV menace-anti-UAV falconry. Vojn Delo 4:146–167 [Internet]. Available from: http://strangevehicles.greyfalcon.us/Nazi%20Pigeon.htm. Cited 5 Aug 2022
89. Brown S (2006) Stealth sharks to patrol the high seas. New Sci [Internet]. Available from: https://www.newscientist.com/article/mg18925416-300-stealth-sharks-to-patrol-the-high-seas/
90. Perry T (2015) Dolphins' and sea lions' important role in guarding against terrorism. Gov Technol [Internet]. Available from: https://www.govtech.com/em/safety/Dolphins-Sea-Lions-Train-Navy-Deployment.html. Cited 10 Dec 2020
91. Treisman R (2022) A Russian naval base is defended by dolphins. It's not as unusual as it sounds [Internet]. National Public Radio. Available from: https://text.npr.org/1095549251. Cited 10 Aug 2022
92. Queremel Milani D, Chauhan P (2024) Genetics, mosaicism [Internet]. StatPearls Publishing, Treasure Island (FL). Available from: https://www.ncbi.nlm.nih.gov/books/NBK559193/
93. Malan V, Vekemans M, Turleau C (2006) Chimera and other fertilization errors. Clin Genet 70:363–373 [Internet]. Available from: https://pubmed.ncbi.nlm.nih.gov/17026615/. Cited 29 Sept 2021
94. De Los Angeles A (2020) Parsing the pluripotency continuum in humans and non-human primates for interspecies chimera generation. Exp Cell Res 387 [Internet]. Available from: https://pubmed.ncbi.nlm.nih.gov/31778671/. Cited 16 Nov 2021
95. Harrison PW, Lower RPJ, Kim NKD, Young JPW (2010) Introducing the bacterial "chromid": not a chromosome, not a plasmid. Trends Microbiol 18:141–148
96. Ito-Harashima S, Mizutani Y, Nishimura M, Kim HJ, Kim YJ, Kim HS et al (2017) A pilot study for construction of a new cadmium-sensing yeast strain carrying a reporter plasmid with the JLP1 promoter. J Toxicol Sci 42:103–109

97. Nagies FSP, Brueckner J, Tria FDK, Martin WF (2020) A spectrum of verticality across genes. PLoS Genet 16:e1009200 [Internet]. Public Library of Science. Available from: https://doi.org/10.1371/journal.pgen.1009200. Cited 23 Apr 2022
98. Brennicke A, Grohmann L, Hiesel R, Knoop V, Schuster W (1993) The mitochondrial genome on its way to the nucleus: different stages of gene transfer in higher plants. FEBS Lett 325:140–145 [Internet]. Available from: https://pubmed.ncbi.nlm.nih.gov/7685713/. Cited 29 Sept 2021
99. Davoodi-Semiromi A, Samson N, Daniell H (2009) The green vaccine: a global strategy to combat infectious and autoimmune diseases. Hum Vaccin 5:488–493
100. Roberts MB, Fishman JA (2021) Immunosuppressive agents and infectious risk in transplantation: managing the "net state of immunosuppression". Clin Infect Dis 73:e1302–e1317 [Internet]. Available from: https://pubmed.ncbi.nlm.nih.gov/32803228/. Cited 20 Apr 2022
101. Sato H, Maharbiz MM (2010) Recent developments in the remote radio control of insect flight. Front Neurosci 4:199
102. Saha D, Mehta D, Altan E, Chandak R, Traner M, Lo R et al (2020) Explosive sensing with insect-based biorobots. Biosens Bioelectron X 6:100050 [Internet]. Available from: https://www.sciencedirect.com/science/article/pii/S2590137020300169
103. He S, Liu S, Li X, Ye M, Wu H, Song J (2025) A review: exploring the designs of bio-bots. Soft Sci 5:5 [Internet]. Available from: https://doi.org/10.20517/ss.2024.50
104. Zhang H-K, Xu B-W, Jia Z-Y, Li B, Feng X-Q (2024) Inverse design of three-dimensional multicellular biobots with target functions. J Mech Phys Solids 187:105634 [Internet]. Available from: https://www.sciencedirect.com/science/article/pii/S0022509624001005
105. Kambouris ME, Manoussopoulos Y, Velegraki A, Patrinos GP (2023) The biote-bot hybrid. The ultimate biothreat merging nanobots, AI-enabled cybernetics and synthetic biology. Future Med AI 1:FMAI4 [Internet]. Available from: https://doi.org/10.2217/fmai-2023-0008
106. Nishi K, Isobe S-I, Zhu Y, Kiyama R (2015) Fluorescence-based bioassays for the detection and evaluation of food materials. Sensors 25831–25867
107. Kambouris ME (2024) Integrating the two: a technical aspect. In: Kambouris ME (ed) Hybrid warfare 2.2: where biothreats meet irregular operations and cyber warriors in the 21st century, 1st edn. Springer Nature, Cham, pp 91–122
108. Kambouris ME (2024) Integrating the two: an operational and strategic aspect. In: Kambouris ME (ed) Hybrid warfare 2.2: where biothreats meet irregular operations and cyber warriors in the 21st century, 1st edn. Springer Nature, Cham, pp 123–139
109. Croddy EA (2005) Weapons of mass destruction volume I: chemical and biological weapons. ABC-Clio, Santa Barbara, CL
110. Smart JK (1997) History of chemical and biological warfare: an American perspective. In: Sidell FR, Takafuji ET, Franz DR (eds) Medical aspects of chemical and biological warfare. Borden Institute, Washington, DC, pp 9–86
111. Spanou G, Daskou M, Kambouris ME, Mitsagga C, Mossialos D, Velegraki A et al (2023) Introducing multifactorial electroculturomics: alternating current electric pulses, combined with mild thermal treatment, exhibit antimicrobial or stimulatory effects on bacterial pathogens and enteroviruses, implying prospects for targeted microbiomics a. bioRxiv [Internet]. Available from: http://biorxiv.org/content/early/2023/11/24/2023.11.24.568545.abstract
112. Hirsch LR, Stafford RJ, Bankson JA, Sershen SR, Rivera B, Price RE et al (2003) Nanoshell-mediated near-infrared thermal therapy of tumors under magnetic resonance guidance. Proc Natl Acad Sci U S A 100:13549–13554 [Internet]. Available from: https://pubmed.ncbi.nlm.nih.gov/14597719/. Cited 5 Dec 2023
113. Gauthier GM (2015) Dimorphism in fungal pathogens of mammals, plants, and insects. In: Goldman WE (ed) PLoS Pathog 11:e1004608 [Internet]. Available from: https://doi.org/10.1371/journal.ppat.1004608. Cited 11 Feb 2019
114. Stathoulias A, Milioni A, Kritikou S, Karmakolia A, Goudoudaki S, Siamoglou S et al (2020) Toward high-throughput fungal electroculturomics and new omics methodologies in 21st-century microbiology and ecology. Omics J Integr Biol 24:493–504

115. Gillings MR (2017) Lateral gene transfer, bacterial genome evolution, and the Anthropocene. Ann N Y Acad Sci 1389:20–36
116. Danchin EGJ (2016) Lateral gene transfer in eukaryotes: tip of the iceberg or of the ice cube. BMC Biol 14:1–3 [Internet]. BioMed Central Ltd. Available from: https://doi.org/10.1186/s12915-016-0330-x. Cited 23 Apr 2022
117. Sharma A, Gupta G, Ahmad T, Krishan K, Kaur B (2020) Next generation agents (synthetic agents): emerging threats and challenges in detection, protection, and decontamination. In: Handbook on biological warfare preparedness. Academic Press, pp 217–256
118. Cieslak TJ, Kortepeter MG, Wojtyk RJ, Jansen HJ, Reyes RA, Smith JO (2018) Beyond the dirty dozen: a proposed methodology for assessing future bioweapon threats. Mil Med 183:e59–e65 [Internet]. Available from: https://pubmed.ncbi.nlm.nih.gov/29401327/. Cited 22 Apr 2022
119. Lai CC, Hsu CK, Yen MY, Lee PI, Ko WC, Hsueh PR (2022) Monkeypox: an emerging global threat during the COVID-19 pandemic. J Microbiol Immunol Infect 55:787–794
120. Khamees A, Awadi S, Al-Shami K, Alkhoun HA, Al-Eitan SF, Alsheikh AM et al (2023) Human monkeypox virus in the shadow of the COVID-19 pandemic. J Infect Public Health 16:1149–1157
121. Farahat RA, Abdelaal A, Shah J, Ghozy S, Sah R, Bonilla-Aldana DK et al (2022) Monkeypox outbreaks during COVID-19 pandemic: are we looking at an independent phenomenon or an overlapping pandemic? Ann Clin Microbiol Antimicrob 21:1–3 [Internet]. BioMed Central Ltd. Available from: https://doi.org/10.1186/s12941-022-00518-2. Cited 1 Dec 2022

Chapter 5
Innovative Aspects of the Updated (Meta)Biothreat and the AI Leverage

Abstract The Meta-Biothreat is not a matter of technologic superiority or miracle-performing means, although these help a lot. But the focus is on the transfer, adaptation and creative use of technology and methodologies developed elsewhere, in other fields, for other purposes, constantly adapting and adjusting assets and means and morphing the milestones for the long-term objectives. The innovative and imaginative use of existing, available and not necessarily high-quality or high-performance means at the perpetrators' disposal in order to maximize effects and achieve objectives is at the center of the Meta-Biothreat, while the AI, although not necessary, may assist in two distinct ways: as an enabler and as a direct or indirect effector. The latter case includes minimally perpetrated bioevents, such as agentless ones, or cyber-enabled and cyber-leveraged ones, which include active bioagents the impact of which is achieved or maximized by—usually cyber-assistance. The former case revolves around different kinds and aspects of logistics, including human resources, materiel and abstract/immaterial assets. The tainting, corruption, spiking or otherwise compromise of bioamenities or the malicious use of valid ones might be a key risk area, as would be the Psychological Operations regarding the prestige and social/international standing of target and alleged perpetrator.

Keywords Geoepidemiology · Surging bioevents · Multifocal bioevents · Recurring outbreaks · Cyber-enabled bioevents · Infodemics · Cyber-leveraged bioevents · Minimally perpetrated bioevents

Exotic bioagents and Synthetic Biology projects do not constitute a Meta-Biothreat. They may assist in its creation and development, but are not essential, nor enough. What is essential is the innovative and imaginative use of resources, existing, known or clandestine, however advanced or not, and this originality is occasionally favored

---

This chapter contains ideas and some excerpts from the paper Disruptive biosecurity: A new concept for preparedness and response to the emerging total biothreat landscape by Kambouris ME et al. (2025) published in Comparative Strategy 1–33, https://doi.org/10.1080/01495933.2025.2565586, further developed beyond the original content.

M. E. Kambouris, *Preparing for the Meta-Biothreat: Disruptive Biosecurity*, Advanced Sciences and Technologies for Security Applications,
https://doi.org/10.1007/978-3-032-25690-4_5

by obsolete means in terms of both delivery and terminal effect. Consequently, exotic and advanced, engineered agents (Chaps. 3 and 4) are not indispensable for the Meta-Biothreat [1], but *may* be one dimension of it. Its essence is out-of-the-box thinking, especially in messing or merging creatively different assets, material and immaterial, which seemingly do not fit together. This is expected to perplex specialists and, even more so, security projection algorithms *and/or* to derail the planned or improvised response efforts. Thus, the essence of the Meta-Biothreat is: (i) The fast turnover; (ii) the dispersed and persistent prerogatives of bioevents; (iii) the AI leverage; (iv) the malicious use/misuse of standard, expanded and novel bioamenities, such as sectors of bioeconomy, green practices and pet propagation and protection; and (v) the high-level disruption and defamation.

5.1 The Surging and Fast Turnover Aspect of Perpetrated Bioevents

The thoughts of Clausewitz regarding concentration of power and striking the enemy center of balance or gravity dominated every aspect of the Western thought, unaffected in essence by the wide distribution of the prerogatives of Sun Tzu within military philosophy and doctrine. It is perhaps for this reason that the concept of the Biothreat, even of perpetrated one, was the very restrictive 3X1: One agent, one location, at one instant/unit of time. There is no reason whatsoever to be so. The modern doctrine of dispersed operations, as apperceived by many international actors, would almost mandate multiple such events, thus denying the spatial Singularity [2]. Then, if the Domino Effect is taken into consideration, the engineering of time intervals that would precipitate if not cause the derailment of the response should be taken into account: The multiplicity of events unfocuses and dissipates the response effort, which may well be combined with the use of more than one agents to further complicate response. There is also the possibility for creative combination: some foci of the delivery being saturated by a single agent each, and some others by a mix. This would plainly revoke the Singularity of agent as well as of Space.

Additionally, if properly timed in succession, or if some events unfold simultaneously and others in succession, the chaos may be maximized and the response nullified: the initial wave would tie down resources, as the Biodefense doctrines do not take into consideration reserves. They are primary and restrictive in scope and mandate to simply build some excessive capacity and stocks, and even this moderate precaution is taken reluctantly, as the financiers are always ready to cut budgets. Thus, when an event is detected, even more so a mix of events, all resources would be earmarked for immediate engagement, more so in cases of highly transmittable diseases, which by definition create (pseudo)multiplicity by dispersion.

In this case there may be some more unsettling issues: new infection foci and new strains, as were with SARS-CoV-2, are usually presupposed to be mutations of some original or cryptic clade of the original strain, which may have been somehow translocated or not and dissipate to new spatial locations [3, 4]. This presupposition refers to spontaneous, naturally occurring events, which is mainstream in public health and medicine. Even if foul play is taken into account, there may be consideration of perpetration or even knowledge of it, but the basic assumption for singular origin, as set for naturally occurring events, remains the basis of the response rationale. Obviously this might not be so, and subsequent releases/strains/variations of the same agent may be separately engineered strains or modified versions of the first construct. It has been rumored that more than one strains of Coronavirus had been modified by GoF experiments, and it may have been presupposed that they were given different functions in the GoF procedure. But this is not a given; in any case the detected SARS-CoV-2 strains were all modified at their spike [5, 6].

In essence, there is a strong possibility, if the Covid-19 has been a perpetrated event, to have different versions of the same pathogen delivered at different space/time intervals. Testing different spatial and chemical versions of the spike is one possible aim. The obvious effect is the fast obsolescence of vaccination and antibody-dependent diagnostic/therapeutic/prophylactic modalities [7] and less so of micromolecular drugs or of biopharmaceuticals other than antibodies/antibody fragments. After some time, perhaps some months or 1–2 years, the original strain may be recycled. In all, the threat as described herein is centered in multiplicity, and it is developing spatiotemporally but also with regard to the bioagents used [8–10]. With the tools of synthetic biology, such massive but diversified efforts of production are within easy reach, especially if low quantities are required, not tons as in industrial production [11]. If many different household or house-brewery-compatible quantities are produced, they would remain under the radar horizon of surveillance systems tuned to detect either massive or spontaneous cases. The former presuppose quantities and bulk in every step of the logistics, the latter showing early signs in a well-developed surveillance system [12, 13]: plants or animals with symptoms; some isolated cases being reported to health providers before the massive wave sets in [14], as in Syndromic Surveillance [1, 15]; or suspect microflora and differentiated microbiomes.

Contrary to the above, perpetrated events would afford minimal advance warning and set in fast. The multiple agents used would eventually shake off suspicions of foul play as it is a deeply ingrained thought that one agent will be used at a time by a perpetrator, especially for Biocrime and Bioterror purposes, rather than a variety. The release of multiple agents has additional advantages to the purely tactical ones discussed above. Even if one agent fails, there is enough contingency potential inbuilt. The agents need not to be thoroughly engineered, or not all of them: an outbreak mobilizes a mechanism that even if operating in dispersed format, it is thoroughly centralized and bureaucratic, and AI assistance in decision making does not change this in essence; but it speeds up decisions and operations. Thus, if alarmed for one pathogen and consequently mobilizing expedient processes for protection, targeted diagnostics/screening, therapy and prophylaxis is focused on this pathogen and the use of agent shifts from the one to another, requiring very different amenities in any of these fields, the whole response is made irrelevant fast. It would have to be rebooted, with great difficulty due to operational inertia. The Monkeypox emergence in the wake of COVID-19 [16, 17] is a brief and comparatively benign foretaste of things to come. If more than one such shifts occur, there would be not enough resources even for the harsh military-style Chinese responses [18, 19]; personnel and instruments will become scarce, and RDTs production surge may not be fast or flexible enough for repeated shifts, as if repeated and dispersed outbreaks occur. This is hardly a criminal scenario, at least in terms of coordination in spatiotemporal terms, but may well be a military or terrorist (domestic or otherwise) operation.

Thus, with limited logistics but an extensive knowledge of Synthetic Biology (or even without the latter), small quantities of a given pathogen may be produced in dispersed locations, thus straining response by spatial occurrence, and exhausting mobilized resources and strain response measures, especially interventive ones like mandatory vaccination. This is achieved more by means of temporal succession than by simultaneous surge, even when one bioagent is used. The use of different, once more easily accessible or harvestable/retrievable agents would further complicate the response effort as there will be a combined threat which is totally unpredictable in its morphing and shifting patterns. This would exhaust the adaptability of response efforts soon: one of the detrimental inheritances of Public Health/Medicinal though in Biosecurity is that, no matter how many agents and ailments an organization prepares against and, cognitively recognizes as risk factors (as, say, in Select Agents Lists [13, 20–23]), its readiness is limited to tackling only *one* at a time. A perpetrator may not abide by this silent protocol.

5.2 The Dispersed and Persistent Prerogatives of Bioevents

As mentioned above, the Multiplicity instead of the Singularity is a key issue in the development of the Meta-Biothreat and it may unfold, inclusively or not, in three planes: spatial (multifocal events), temporal (recurring or sequential events) and microbiological (multi-agent bioevents) [2]. This kind of perpetration is suitable for basic or advanced threats in terms of mechanisms and pipelines. Its most basic but still non-Singularity application would be the easiest to deploy and the simplest, though not easy, to respond to. It is a quasi-singularity, insinuating just one agent, which greatly simplifies logistics, slashes expenses, effort and exposure of the perpetrator(s) and still may be extremely tasking for well-equipped and prepared response agencies and conglomerates which, however, retain a restrictive, Public Health—inspired apprehension of single threats. Not taking into account batch samples, meaning similar but different strains as, for example, allegedly created in Wuhan, for experimental purposes of course [24, 25], nor the secondary foci -in spatiotemporal and objective terms—where spontaneous evolution and adaptation of the initial agent may have occurred, the plain use of different aliquots of the same preparation/strain may be released in multiple foci with or without time intervals. It is a simple issue to achieve release in different locations with plain and inexpensive means and thus create multi-focus outbreaks, synchronous or asynchronous. This would strain resources by the mere need to address and contain occurrence in different locations, leading in failure location-specific measures, as the exclusion and quarantine zones seen in movies. A general lockdown would be seemingly necessary. The need to cover distinct and widely dispersed localities and a wide population basis with protection, prophylaxis, screening and comprehensive diagnostics and therapeutic intervention, in terms of care, medical attention and medical countermeasures (MCM) such as therapeutic agents and life support, is tasking to the extreme [26].

In many cases connectivity alleviates problems, as with panels of experts being centrally located and being able to assess data collected by on-site, stationed or expedited teams with suitable, portable, RDT and PoN and, better still, RLS-compatible instruments and equipment. But dispatches, even if "droned" in and out (which entails a number of drone-specific dangers) is a limited commodity, especially if under lockdown. Multiple rather than a unique, widespread lockdowns create regulatory implications in many levels. One of these is the prospective accumulated multi-factorial damage of the carpet quarantining of large areas or in a nationwide fashion, which has been a preferred precautionary measure, perhaps prompted by a number of widely dispersed outbreak foci that would create the disruptive impression of a state-wide epidemic and trigger precautionary rather than reactionary reflexes of panels of experts [27].

Such quasi-singular, sole-agent outbreaks may be simple, plain and not particularly destructive—perhaps even unimpressive, but after the COVID-19 few responsible panels and individual decision-makers would take the risk. Thus, although moderately destructive, they may prove extremely disruptive and pave the way for other hostile acts, such as internal dissension and external aggression. Given their implementational simplicity and affordability [28], *sensu lato*, such events may be repeated, thus qualifying for classification as recurring outbreaks. Still, shortly after they are resolved and there has been a return to normality, but with the impressions still very much alive and during the effort to recuperate from all damage due to shutting down a modern economy based on uninterrupted activity and operations, an identical series of outbreaks may occur. In such a case the consequences are hard to predict. Having returned to normality means that everything deployed under duress is retrieved: the personnel laid off or returned to normal stations and the materiel repositioned to normality, or discarded, if unserviceable or unretrievable, or put in storage. It is not difficult to create a follow on, identical incident, and somewhat improved, due to lessons learnt from the first time: no new agent, no new methodology. Just another wave, as the waves of the Plague in the Middle Ages or the waves of the Plague of Athens; Epidemiology is very much aware of waves, and even COVID-19 was coming in season-dependent waves. A new wave would require one of the same: the authorities, no matter their effectiveness in the first round would necessarily have over-bragged over their performance, to instill trust and avoid audits for possible hick-ups, corruption, misdemeanor or plain incompetence, which in such conditions cost millions in currency and many lives. There is null possibility not to respond to the same stimulus, even if of patently low impact; the Public would demand so, the inertia of social mechanics would prompt such thoughts and authorities would need to improve their performance, showing that they have learnt their lessons [29, 30].

Still, the psychological, not to mention the disruptive blow would be tremendous: it was the 2023 wave of the COVID-19, after a seasonal break in Summer 2022, that caused the most social dissention and lack of trust, even before the issues of violently enforced vaccination [31]. There was a sense of vainness for the previous efforts and sacrifices, a decline in trust to experts and their prescribed responses as they could not avert another round of a now known threat [32]. The repeated, recurring event is much more disruptive and accumulatively destructive than a unique one, even if the latter was to be more destructive in nature. It is the intermission, the relaxation and then once more the stimulus that create a sequence which shatters morale but also resilience in terms of organisms and systems [33].

Such approaches are the quintessence of the Meta-Biothreat. As obvious, it is not a matter of technologic superiority or miracle-means, although these help a lot. But the focus is on the *use* of technology and methodology, possibly developed elsewhere, for other purposes; on constantly adapting means and morphing milestones in the long-term objectives; and on imaginatively using existing, not necessarily high-quality or performance means at the perpetrators' disposal to maximize effect (for example delivery of bioagents by recreational drones). After the quasi-conventional but actually deeply hybrid exchanges of the 2022–2025 timeframe between Russia and Ukraine [34], such practices and their psychological impact must be mulled upon, appreciated, assessed and incorporated in Biosecurity schemes to disrupt the pathway of realization, as will be hinted in Chaps. 6–8).

5.3 The AI Leverage

The AI may assist a bioperpetrator in two distinct ways: as an enabler and as a direct effector. Many of the risk areas are covered in this chapter, like the perpetrated failure of medical tests and life support instruments, the hostilization of company and wildlife animals, the industrial and home incidents, especially "smart home" mal- and dys-functions. Virtual companions and assistants that may be turned physically and/or digitally hostile, but much more so through hidden AI functionality or under AI guidance should not escape mention. The combination of Industry 4.0, Internet-of-Things and 5G Internet allows an unprecedented degree or reach and infiltration to the most hidden corners of the environment, especially urban, but perhaps rural too, while satellite transmission covers much of the wildlife/untainted surface environment, like the wilderness-jungles, steppes, desert and mountains.

5.3.1 Direct-Effecting: Infodemics

The Infodemics are ill-defined [35], and more so the minimally perpetrated bioevents; but in here the second term is used as an umbrella concept for the agentless (bio)events, the cyber-leveraged and the cyber-enabled/perpetrated bioevents (as defined and discussed in Chap. 6); the first term, on the other hand, is more restrictive and refers to the two first cases, but not the third.

The agentless events are to be enhanced by active use of AI components. It could facilitate production of fake news and images, replicate or disable or mimic news traffic and live broadcasts and jam response messaging through software-controlled or -defined communication facilities, thus creating a public health hazard in the minds and daily processes of the public and even of susceptible government agencies without one actually existing. The existence, progress, development and impact of an outbreak is apperceived and managed through remote communications and software-controlled instruments [36], especially if quarantine, shutdowns, lockdowns and other suppressive measures are implemented, more so if implemented by algorithms or automatically. If these are corrupted or taken over, all the metrics, markers and procedures of epidemics may be replicated, not to mention false patients (real patients with falsified records and fake emergency calls reporting fake symptoms and then fake reports from the responders). If prescription is taken into account there would be not only a disruptive effect of catastrophic proportions, but also a destructive one due to fake prescription. All these may or might be implemented by conventional hacking, and amplified through Big Data processing and 5G Internet and IoT [37, 38], but the AI makes it much more impactful [1].

The Cyber-leveraged events are a step before the agentless events in the evolutionary threat charts: a bioevent of any magnitude but of spreadable nature (through common source or transmittable in nature) that would be of minor importance, causing little if any disruptive effect and perhaps fully curable in nature, may be amplified out of proportions. This is not as in the previous case, whence by creating an *impression* of the bioevent and its results detrimental, adverse or outright disastrous outcomes might be engineered. In this iteration actual, tangible bioresults are effected. Jamming or falsifying reports, guidelines, instructions to public and to responders, false prescriptions and engineered falsified diagnostic equipment reports would result in much larger spread of the disease, inefficient or null response and massive health casualties, fatal or not. Similarly but on a different context, corrupting normal procedures and processes of life- or health-support equipment, such as pacemakers, reactive or timer-activated pumps and other delivery amenities, like the new, actively controlled diadermal delivery smart patches with close-loop decision-making [39], but also standard prescription for chronic or incidental illnesses are just as threatening prospects. Simple hacking would do enormous damage, but AI-initiation would take it to new levels, for example by timing results with some major social event or by targeting key officials within some narrow but vital temporal window.

The most basic is, of course, the cyber-causality of a true bioevent, i.e. the cyber-enabled bioevents. In this case there is an actual bioevent, of whatever magnitude, location and nature, but it has not occurred spontaneously. It falls within the perpetrated bioaccidents: the Sverdlovsk Anthrax leak incident of 1979 in the-then USSR is a major and characteristic accidental such occurrence [40], but its spontaneity may be debated. Thus, the perpetrated bioaccident is not a risk entity solely associated with cyber applications; plainer methods may be used. But in the case examined herein it may be *engineered* remotely, by manipulating electronic safeguards and automated processes, either of robotic execution, or of message-defined human execution; for example a direct order to misplace or deliver a lethal agent, and subsequent falsification of voice and data contacts requesting confirmation or expressing objections [1]. Similarly, mislabeling of agent cultures, and failure of containment equipment and surveillance instruments may be engineered. It must be reminded that supposedly two strains of Smallpox exist (at least officially), both under strict guard and security [41]. A perpetrated accident could initiate a Smallpox epidemic at a timeframe when no doctor has any idea of its symptoms and cure, no public health official of its actual epidemiological dynamics in naïve and dense populations that travel daily and in intercontinental or planetary distances. There are many detrimental factors if such an event occurred: no diagnostics have been field-tested, no respective response protocols activated nor vaccination and immune memory residing. Similarly, one of the official storylines of the COVID-19 pandemic is the accidental release (quotation marks on *accidental* may be appropriate) from an elaborate lab facility [5], without any elaboration of the security/safety pitfalls that allowed such catastrophic an event.

5.3.2 Enabling-Facilitating of Actual Perpetration

The AI is a unique enabler for issues needing plain but extensive kinds of expertise as with different (but not all) aspects of logistics for perpetration of Biothreats. Regarding materiel, it may create shopping lists for reagents, instrumentation, expendables and glassware; implement the acquisition of said items and relative services (delivery, other services such as installation-attuning) by funneling credits and cash in highly complicated pipelines to evade tracking by the authorities and money laundering aversion initiatives; and then suggest delivery schemes and addresses to bypass authorities. Last, but not least it may also suggest alternative items of higher availability or commoner, more affordable and less surveyed and controlled standard laboratory items but also bioagents of interest for a given task [20, 42]. AI may also facilitate financing by covert or fake-cause fundraisings, identifying possible sponsors and suggesting approach strategies.

The human resources aspect of the logistics is also a prominent field for AI amenities. Videos, images and texts for indoctrination and recruitment, based on reality or sous-reality may be combined with selection of prospective recruits by their interests in the cyberspace. Interested individuals may be filtered by successive virtual interviews and the ones selected they may take AI-advised initial, introductory and even advanced courses, attend advanced indoctrination courses, and later participate, if needed, in 5G-transmitted workshops of enhanced virtual reality [1, 42]. If, as claimed, "with 5G Internet something happening somewhere happens everywhere" due to the realism of the recreation/ recording/transmission, this applies also to advanced courses in microbiology, on how to acquire and grow bioagents and in biotechnology, on weaponizing and delivering/dispersing them. Finally, successful recruits may be verified through gaming and simulation rounds to improve their skills and improve planning, while operational planning and selection of the optimum recruits and means from the available respective pools may be AI defined or assisted as well [43].

5.4 The Malicious (Mis)Use of Bioamenities

5.4.1 Biologicals/Biopharmaceuticals/Microbials/Probiotics/ Pharmabiotics

It is debatable whether the use of synthetic therapeutic compounds is bound to be superseded by biologicals of every kind. These range from probiotics, which are edibles that promote proper physiological functionality and host-microbiome interface, and pharmabiotics, which additionally secrete active therapeutic moieties, including but not limited to natural, unprocessed and unmodified antibiotics, to biopharmaceuticals. The last ones are preparations of live agents or their products with specific therapeutic value, and possibly prophylactic applicability, which are taken by prescription; for any length of time, but finite. The pharmabiotics may be considered rather as supplements of pharmacological activity than pharmaceuticals for acute or chronic intervention [44–47].

It is obvious that such preparations may become platforms for nefarious bio-offensive purposes. They may contain pathogenic strains that would enter through the nominated entry point and method of the preparation, which is less complicated and more direct approach, likely to bring acute results, not long-term ones as does the failure (or rather denial) of therapy—although therapies might occasionally well bring short-term or even, conceivably, immediate results if failing or compromised [48]. But the denial of therapy needs far less offensive agent to be introduced into the preparations than the active infection/intoxication and may be effected by inclusion of aggressive strains that would deplete the beneficial microbiota or molecules. Acts of sabotage are best served by the second method, massive attacks for strategic, long-term results by the first, regardless of the number of the target population size. The most elaborate scheme is the inclusion in the active preparation, before or conceivably even after its administration, of transformation agents (transformation taken *sensu lato*), that would mutate the beneficial or benign microbial compounds to virulent ones, obviously through some sort of LGT [49]. The more efficient method is some scheme based on bacterial conjugation and plasmid exchange/dissipation through the microbial population. The simplest would be based on transformation (*sensu stricto*) by naked DNA; its applicability is conditional however, dependent on the nature of the specific preparation. The most robust would be the corruption of the benevolent strains through viral transduction, as virions are very stable and survivable, especially virions with double-stranded DNA. It is obvious that only transduction *may be* applicable to viral pharmabiotics/biopharmaceuticals, such as the virotherapeutic preparations, and this due to the advent of virophages [50, 51].

5.4.2 Industrial Strains

The potency of industrial strains may make them highly detrimental for living hosts and materiel, as they are efficient and hardy [52–54]. The simple extraction/acquisition of strains of specific qualities may be enough once delivery schemes are sorted out. But less hazardous strains may be opted for, which will be less closely guarded in the context of any industry of whatever size (especially small-medium units, however, as these are more sustainable). Being efficient and hardy, they may be used as the end platform, by inserting determinants of virulence, or enzymes for non-virulent terminal effect, like corrosion and degradation of plastic, fuel, linings of optics or rusting of metals. Alternatively, they may be used as source for traits to ruggedize existing (or engineered) pathogens with high virulence but lacking in persistence, efficiency, resistance, fastidiousness or survivability and stability. Any method of LGT [49, 55] would assist both such options, and spiking or substitution of foodborne or food industry microorganisms with some selected industrial species may cause large quantities of foodstuff to become poisonous and thus result in health casualties of humans, animals or even microbial populations used productively or in regulatory capacities, and perhaps causing food and drug shortages and insecurity. Another such approach would be to destroy the bioindustrial production of some facility by mutating the starting cultures with toxic genes or detrimental elements (enzymes, toxins) degrading or corrupting the raw materials or the products, rather than simply ceasing the production process.

5.4.3 Pets, Wildlife and Bred Animals

The interface of animals and humans allows animal pathogens to cross species barriers and infect naïve human populations, as has happened with the HIV [56] and the SARS-CoV-2, at least allegedly [57, 58], not to mention the Plague that disseminates by fleas hosting *Y. pestis* and hosted, themselves, by rodents from which they shift to humans and deliver the pathogen [59]. The Rabies is yet another example of an animal pathogen (virus) crossing from one animal species to another [60], *B. anthracis* usually crosses from livestock to farmhands [61] and fungi cross from plants to animal hosts, the Farmer's Lung being a known example [62]. As a result, animal populations interfacing deeply with human hosts, from pigeon species at terraces, plazas and national monuments to company animals in backyards and apartments, may become highways for infection by a diverse number of bioagents.

The threat becomes trickier by the fact of massive chipping (RFID-Radio Frequency IDentification) such populations, from the mandatory procedure to company animals to the research such practices in Wildlife studies [63]. Elaborate scenarios may develop: the uncomplicated one is such animals through neurohacking [64] to be ordered to attack specific targets or run amok and attack fully randomly, or a without obvious selectivity. More complex approaches would be to reach some reservoir of a pathogen, possibly a rabid animal, and after becoming vectors or fomites themselves, to initiate dissemination routines, from attacking (to transmit Rabies) to simply coming into contact (to deliver fleas). Given that neurohacking has allowed control of insects' flight by radio control [65], the transition to remotely controlling animals, as sensors and as effectors, through screens and remote controls or through interfaces currently used for First Person View (FPV) drones is not science fiction. But it does remains extremely elaborate and thus applicable by a preciously low number of players in possession of such technology and expertise, state or non-state ones.

5.5 High-Level Disruption and Defamation

5.5.1 Deniability

Depending on the strategic and even the political environment and the nature and identity of the prospective perpetrator, in some cases subtlety and discretion are essentials for meeting the objectives without retribution; thus deniability is occasionally a major prerogative in some operations. In the opening months of the Ukrainian conflict, the Head of the Russian equivalent of the NBC warfare forces came up with allegations of US-Ukrainian plans for bioattacks with agents and methods so as to blend to local geoepidemiology patterns [66] and the migration patterns of vector wildlife species, namely birds, obviously carrying and disseminating the Avian Flu, allegedly of some selected or enhanced, highly virulent strain,[1] so as to maintain deniability and to derail surveillance, response and retribution, while maximizing effectiveness.

5.5.2 Disinformation, False Flag and the Blame Game

The terror inbred within human nefarious, million-years-old past for infectious diseases and health ailments in general makes the use, alleged use or threat of use of bioagents a true Psyops weapon of great potential and amenable to clever wielding. The fake news on Iraq's alleged BW program led to the 2003 multinational invasion and destruction of the said polity [67, 68], while, despite the perceived instability and unreliability [41, 69, 70], the deterrence inherent in the mere production and stocking of BW is very high: reliable or not, no-one wants to risk infection of urban and rural environments to the point of inhabitability, as occurred at the Anthrax-infected (or infested) Gruinard island of the UK due to testing of such weapons during WW II

[1] https://www.globaltimes.cn/page/202203/1254528.shtm.

[61, 68, 71]. On the other hand, subtle uses are unproven but rumored/alleged factual events, for state-sponsored and domestic bioterrorism [72, 73] and different versions of crime [74–77].

As a result it might be educative to explore the alleged Chinese governmental implication in the premeditated or opportunistic exploitation of the pandemic nature of the COVID-19, a fact that could perplex the international relations of the country and implicate officials in international criminal cases. The whole argumentation line that the Chinese had purposely done *anything* to release the Covid-19 plague and harm the West is unconvincing, contrary to the view of a wide audience and opinion-makers (genuine or to upkeep appearances). The Legalistic school of thought, followed by the current leadership of China [78], would discourage such proceedings in method, propriety and engineering, and perhaps objectives too.

Actually, China had little to gain from a pandemic—it hurt China most of all [79, 80]. On the other hand this event would provide an opportunity to the West to enter, once more, in "Crisis mode", enacting a modified Diversionary War as described by the Diversionary War Theory-DWT [81]. It is obvious that after the end of Cold War pleasantries, the West seeks some Evil to fight against. The War on Drugs, the War on Terror and the current millennium successive epidemics and climate crisis underscore the argument [25] and could qualify for a Diversionary Crisis concept, rather than Diversionary Wars and much subtler than the latter.

Further, there were no signs of exploitation by the Chinese regarding disinformation or defamation: it was American officials calling it "China Virus" [81, 82]. The Chinese are secretive by nature and much more so when their prestige is to take a blow. The same with the Japanese, who lost a war by not torpedoing merchantmen [83], for reasons that might ultimately relate to considering such targets beneath their Samurai- inherited attitude and decency. The Chinese might have stalled in providing data, but not through malevolence; they displayed cold temper and lack of empathy/ indifference, but this is something else and definitely not a Chinese-only trait. Cover-ups to great public damage is not something unheard of in the First World, not to mention the long-departed Second World; so much so that the Conspiracy Theories are a pastime favored by an increasing part of the population in the former—not without reason.

It is not the Chinese that made the WHO not to declare a pandemic. Despite any degree of infiltration, they possess no such leverage if the other major players, especially the US, had a firm opposite stance. Seeing the elimination approach or Zero-COVID policy with brutal, militarized crackdowns in Chinese cities [84–86], their last priority would be to use the pandemic so as to undermine the West. The West, however, might have been implicated in the Chinese origin of the pandemic, by funding GoF experiments in Wuhan [25]. With or without previously engineering a release in Wuhan, there were Western officials who declared in favor of requesting indemnities for the staggering economic losses [87]—similar to what had happened with the Opium Wars [88].

References

1. Kambouris ME, Manoussopoulos Y, Velegraki A, Patrinos GP (2025) Disruptive biosecurity: a new concept for preparedness and response to the emerging total biothreat landscape. Comp Strateg 1–33 [Internet]. https://doi.org/10.1080/01495933.2025.2565586
2. Kambouris ME (2023) Global catastrophic biological risks in the post-COVID-19 world: time to act is now. OMICS 27:153–170 [Internet]. Available from: https://pubmed.ncbi.nlm.nih.gov/36946656/. Cited 24 Jul 2023
3. Chiara M, Horner DS, Gissi C, Pesole G (2021) Comparative genomics reveals early emergence and biased spatiotemporal distribution of SARS-CoV-2. Mol Biol Evol 38:2547–2565 [Internet]. Available from: https://pubmed.ncbi.nlm.nih.gov/33605421/. Cited 1 Dec 2022
4. Poorolajal J (2020) Geographical distribution of COVID-19 cases and deaths worldwide. J Res Health Sci 20:e00483
5. Holmes EC, Goldstein SA, Rasmussen AL, Robertson DL, Crits-Christoph A, Wertheim JO et al (2021) The origins of SARS-CoV-2: a critical review. Cell 184:4848–4856 [Internet]. Available from: /pmc/articles/PMC8373617/. Cited 5 Aug 2022
6. Magazine N, Zhang T, Wu Y, McGee MC, Veggiani G, Huang W (2022) Mutations and evolution of the SARS-CoV-2 spike protein. Viruses 14
7. Rabaan AA, Al-Ahmed SH, Albayat H, Alwarthan S, Alhajri M, Najim MA et al (2023) Variants of SARS-CoV-2: influences on the vaccines' effectiveness and possible strategies to overcome their consequences. Medicina (Kaunas) 59
8. Kambouris ME, Gaitanis G, Manoussopoulos Y, Arabatzis M, Kantzanou M, Kostis GD et al (2018) Humanome versus microbiome: games of dominance and pan-biosurveillance in the Omics universe. Omi A J Integr Biol 22:528–538
9. Kambouris M (2021) The concept of humanome and the microbiomic dimension. In: Kambouris M (ed) Genomics in biosecurity, 1st edn. Elsevier Academic Press, London, pp 15–28
10. Kambouris ME, Manoussopoulos Y, Kantzanou M, Velegraki A, Gaitanis G, Arabatzis M et al (2018) Rebooting bioresilience: a multi-OMICS approach to tackle global catastrophic biological risks and next-generation biothreats. Omi A J Integr Biol 22:35–51
11. Zilinskas R (2016) The Soviet biological weapons program and its legacy in today's Russia. Washington DC. Report No. 11
12. Kambouris ME (2018) Mobile stand-off and stand-in surveillance against biowarfare and bioterrorism agents. In: Karampelas P, Bourlai T (eds) Advanced sciences and technologies for security applications. Springer, pp 241–255
13. Leiser OP, Hobbs EC, Sims AC, Korch GW, Taylor KL (2021) Beyond the list: bioagent-agnostic signatures could enable a more flexible and resilient biodefense posture than an approach based on priority agent lists alone. Pathogens 10:1497 [Internet]. Multidisciplinary Digital Publishing Institute. Available from: https://www.mdpi.com/2076-0817/10/11/1497/htm. Cited 6 Oct 2022
14. Sharma A, Farouk IA, Lal SK, Martinez-Sobrido L, Toral FA (2021) COVID-19: a review on the novel coronavirus disease evolution, transmission, detection, control and prevention. Viruses 13:202 [Internet]. Multidisciplinary Digital Publishing Institute. Available from: https://www.mdpi.com/1999-4915/13/2/202/htm. Cited 20 Apr 2022
15. Velsko S, Bates T (2016) A conceptual architecture for national biosurveillance: moving beyond situational awareness to enable digital detection of emerging threats. Heal Secur 14:189–201 [Internet]. Available from: http://www.ncbi.nlm.nih.gov/pubmed/27314659. Cited 14 Oct 2019
16. Farahat RA, Abdelaal A, Shah J, Ghozy S, Sah R, Bonilla-Aldana DK, et al (2022) Monkeypox outbreaks during COVID-19 pandemic: are we looking at an independent phenomenon or an overlapping pandemic? Ann Clin Microbiol Antimicrob 21:1–3 [Internet]. BioMed Central Ltd. Available from: https://ann-clinmicrob.biomedcentral.com/articles/10.1186/s12941-022-00518-2. Cited 1 Dec 2022

17. Khamees A, Awadi S, Al-Shami K, Alkhoun HA, Al-Eitan SF, Alsheikh AM et al (2023) Human monkeypox virus in the shadow of the COVID-19 pandemic. J Infect Public Health 16:1149–1157
18. Kania E, McCaslin I (2020) People's warfare against COVID-19: testing China's military medical and defense mobilization capabilities [Internet]. Institute for the Study of War, Washington DC. Available from: https://www.jstor.org/stable/resrep27524?searchText=&searchUri=&ab_segments=&refreqid=&searchKey=
19. Bacchus P, Nissen K, Berg J, Bråve A, Gyll J, Larsson C et al (2021) Civil-military collaboration to facilitate rapid deployment of a mobile laboratory in early response to COVID-19: a high-readiness exercise. Heal Secur 19:488–497
20. Kambouris ME, Velegraki A, Patrinos GP, Poulas K (2014) Reviewing the US select agents and toxins list with more efficient assessment of the potential applicability of microbes in bioterrorism: the fungal agent. Am J Bioterror Biosecur Biodefense 1:3–5 [Internet]. Available from: www.austinpublishinggroup.com. Cited 15 Oct 2019
21. Casadevall A, Imperiale MJ (2019) Destruction of microbial collections in response to select agent and toxin list regulations. Biosecur Bioterror 8:151–154 [Internet]. Available from: http://www.liebertpub.com/doi/10.1089/bsp.2010.0012. Cited 22 Mar 2019
22. Kambouris ME (2024) The GCBR after COVID-19. Hybrid Warf 22, 1st edn. Springer, Cham, pp 143–186
23. Siatis KE, Vonitsanos G, Kanavos A, Manoussopoulos Y, Velegraki A, Patrinos GP et al (2024) Smintheus DB: a publicly accessible amenity for informing, regulating, risk-managing and deliberating on current and retrospective biosecurity research with minimal IT and cognitive requirements. JOJ Pub Heal 9:555761
24. Gostin L, Gronvall G (2023) The origins of COVID-19—why it matters (and why it doesn't). N Engl J Med 388:2305–2308 [Internet]. https://doi.org/10.1056/NEJMp2305081
25. Berche P (2023) Gain-of-function and origin of Covid19. Presse Med 52:104167
26. Vardavas C, Zisis K, Nikitara K, Lagou I, Marou V, Aslanoglou K et al (2023) Cost of the COVID-19 pandemic versus the cost-effectiveness of mitigation strategies in EU/UK/OECD: a systematic review. BMJ Open 13:e077602. Available from: http://bmjopen.bmj.com/content/13/10/e077602.abstract
27. Gibson-Fall F (2021) Military responses to COVID-19, emerging trends in global civil-military engagements. Rev Int Stud 47:155–170 [Internet]. Cambridge University Press. Available from: https://www.cambridge.org/core/journals/review-of-international-studies/article/military-responses-to-covid19-emerging-trends-in-global-civilmilitary-engagements/1A68DD0D2F22C5081507FD15743FF2AD. Cited 7 Sep 2023
28. Croddy E, Krčálová S (2001) Tularemia, biological warfare, and the battle for Stalingrad (1942–1943). Mil Med 837–838 [Internet]. Association of Military Surgeons of the US. Available from: https://academic.oup.com/milmed/article/166/10/837/4819330. Cited 28 Sep 2020
29. Assefa Y, Gilks CF, Reid S, van de Pas R, Gete DG, Van Damme W (2022) Analysis of the COVID-19 pandemic: lessons towards a more effective response to public health emergencies. Global Health 18:1–13 [Internet]. BioMed Central Ltd. Available from: https://globalizationandhealth.biomedcentral.com/articles/10.1186/s12992-022-00805-9. Cited 7 Sep 2023
30. Aruru M, Truong HA, Clark S (2021) Pharmacy emergency preparedness and response (PEPR): a proposed framework for expanding pharmacy professionals' roles and contributions to emergency preparedness and response during the COVID-19 pandemic and beyond. Res Soc Adm Pharm 17:1967–1977
31. Bence Farkas C, Gábor Dudás C, Gergely Csaba Babinszky C, Lászí Földi C (2023) Analysis of the virus SARS-CoV-2 as a potential bioweapon in light of international literature. Mil Med 188:531–540. https://doi.org/10.1093/milmed/usac123
32. Adam S, Urman A, Arlt D, Gil-Lopez T, Makhortykh M, Maier M (2023) Media trust and the COVID-19 pandemic: an analysis of short-termtrust changes, their ideological drivers and consequences in Switzerland. Commun Res 50:205 [Internet]. SAGE Publications. Available from: /pmc/articles/PMC9663272/. Cited 7 Sep 2023

33. Ludovici G, Gabbarini V, Cenciarelli O, Malizia A, Tamburrini A, Pietropaoli S et al (2016) A review of techniques for the detection of biological warfare agents. Science and Technology Research Institute for Defence 8:17–26
34. Bosneagu R (2024) Russia-Ukraine war of drones—strategic impact, tactics, and implications. The drones dropped in Romania during the Russia-Ukraine war. Rom Mil Think 2024:504–521
35. Borges do Nascimento IJ, Pizarro AB, Almeida JM, Azzopardi-Muscat N, Gonçalves MA, Björklund M et al (2022) Infodemics and health misinformation: a systematic review of reviews. Bull World Health Organ 100:544–561
36. Chatzis I, Rompolas P, Karachristos C, Grivopoulou A (2023) Cyberbiosecurity: the threat of the sages. In: Kambouris ME (ed) Biosecurity in the making: the threats, the aspects and the challenge of readiness, 1st edn. CRC Press, Boca Raton, pp 119–132
37. Özdemir V (2018) The dark side of the moon: the Internet of Things, Industry 4.0, and the quantified planet. Omi A J Integr Biol 22:637–641 [Internet]. Available from: https://www.liebertpub.com/doi/10.1089/omi.2018.0143. Cited 2019 May 10
38. Yaacoub JPA, Noura HN, Salman O, Chehab A (2023) Ethical hacking for IoT: security issues, challenges, solutions and recommendations. Internet Things Cyber-Physical Syst 3:280–308
39. Liang S, Liu S, Long Z, Li X, Han Q, Wei X et al (2025) A self-powered hydration-monitoring and drug-delivery skin patch for closed-loop treatment of atopic dermatitis. Microsystems Nanoeng 11:156. https://doi.org/10.1038/s41378-025-01000-7
40. Sahl JW, Pearson T, Okinaka R, Schupp JM, Gillece JD, Heaton H et al (2016) A *Bacillus anthracis* genome sequence from the Sverdlovsk 1979 autopsy specimens. MBio 7 [Internet]. American Society for Microbiology. Available from: https://pubmed.ncbi.nlm.nih.gov/27677796/. Cited 3 Nov 2020
41. Croddy EA (2005) Weapons of mass destruction volume I: chemical and biological weapons. ABC-Clio, Santa Barbara, CL
42. Sandbrink JB (2023) Artificial intelligence and biological misuse: differentiating risks of language models and biological design tools. Available from: https://arxiv.org/abs/2306.13952v5. Cited 30 Oct 2023
43. Caldwell M, Andrews JTA, Tanay T, Griffin LD (2020) AI-enabled future crime. Crime Sci 9:14. https://doi.org/10.1186/s40163-020-00123-8
44. Declerck PJ (2012) Biologicals and biosimilars: a review of the science and its implications. Generics Biosimilars Initiat J Pro Pharma Commun Int 1:13–16
45. Spadiut O, Capone S, Krainer F, Glieder A, Herwig C (2014) Microbials for the production of monoclonal antibodies and antibody fragments. Trends Biotechnol 32:54 [Internet]. Elsevier. Available from: /pmc/articles/PMC3906537/. Cited 30 Nov 2022
46. Kooijman M, van Meer PJK, Moors EHM, Schellekens H (2012) Thirty years of preclinical safety evaluation of biopharmaceuticals: did scientific progress lead to appropriate regulatory guidance? Expert Opin Drug Saf 11:797–801 [Internet]. Available from: http://www.ncbi.nlm.nih.gov/pubmed/22861668. Cited 10 May 2019
47. Mantle JL, Rammohan J, Romantseva EF, Welch JT, Kauffman LR, McCarthy J et al (2019) Cyberbiosecurity for biopharmaceutical products. Front Bioeng Biotechnol 7:116
48. Noronha V, Joshi A, Patil VM, Bhosale B, Muddu VK, Banavali S, et al (2014) Pattern of infection, therapy, outcome and risk stratification of patients with febrile neutropenia in a tertiary care oncology hospital in India. Indian J Cancer 51. Available from: https://journals.lww.com/indianjcancer/fulltext/2014/51040/pattern_of_infection,_therapy,_outcome_and_risk.31.aspx
49. Stokes HW, Gillings MR (2011) Gene flow, mobile genetic elements and the recruitment of antibiotic resistance genes into gram-negative pathogens [Internet]. FEMS Microbiol Rev 790–819. Oxford Academic. Available from: https://academic.oup.com/femsre/article/35/5/790/2680373. Cited 2020 Sep 23
50. Nettelbeck DM, Leber MF, Altomonte J, Angelova A, Beil J, Berchtold S, et al (2021) Virotherapy in Germany—recent activities in virus engineering, preclinical development, and clinical studies. Viruses 13 [Internet]. Multidisciplinary Digital Publishing Institute (MDPI). Available from: /pmc/articles/PMC8402873/. Cited 20 Aug 2022

51. Barrow P, Dujardin JC, Fasel N, Greenwood AD, Osterrieder K, Lomonossoff G, et al (2020) Viruses of protozoan parasites and viral therapy: is the time now right? Virol J 17 [Internet]. Available from: https://pubmed.ncbi.nlm.nih.gov/32993724/. Cited 20 Aug 2022
52. Bustamante-Torres M, Romero-Fierro D, Estrella-Nuñez J, Cuadros-Buenaventura E, Bucio E (2022) Application of microbes in biotechnology, industry, and medical field BT—application of microbes in environmental and microbial biotechnology (Inamuddin, Ahamed MI, Prasad R, eds). Springer Nature Singapore, Singapore, pp 113–152. https://doi.org/10.1007/978-981-16-2225-0_4
53. Budanova E V, Sheina NI, Skriabina EG, Ivanov NG, Vorob'ev AA, Zverev VV (2010) Comparative studies of microbial strains used in biotechnology for the elaboration of biosafety standards. Vestn Ross Akad meditsinskikh Nauk. Russia (Federation) 42–46
54. La Maestra S, D'Agostini F, Geretto M, Micale R (2021) Microbial-based cleaning products as a potential risk to human health: a review. Toxicol Lett 353
55. Danchin EGJ (2016) Lateral gene transfer in eukaryotes: tip of the iceberg or of the ice cube. BMC Biol 14:1–3 [Internet]. BioMed Central Ltd. Available from: https://bmcbiol.biomedcentral.com/articles/10.1186/s12915-016-0330-x. Cited 23 Apr 2022
56. Van Baarlen P, Van Belkum A, Summerbell RC, Crous PW, Thomma BPHJ (2007) Molecular mechanisms of pathogenicity: how do pathogenic microorganisms develop cross-kingdom host jumps? FEMS Microbiol Rev 31:239–277 [Internet]. Available from: https://pubmed.ncbi.nlm.nih.gov/17326816/. Cited 30 Apr 2022
57. Steele EJ, Gorczynski RM, Lindley RA, Tokoro G, Temple R, Wickramasinghe NC (2020) Origin of new emergent coronavirus and Candida fungal diseases-terrestrial or cosmic? Adv Genet 106:75–100 [Internet]. Elsevier. Available from: https://pubmed.ncbi.nlm.nih.gov/33081928/. Cited 29 Nov 2022
58. Lu G, Wang Q, Gao GF (2015) Bat-to-human: spike features determining "host jump" of coronaviruses SARS-CoV, MERS-CoV, and beyond. Trends Microbiol 23:468–478 [Internet]. Available from: https://pubmed.ncbi.nlm.nih.gov/26206723/. Cited 30 Nov 2022
59. Demeure C, Dussurget O, Mas Fiol G, Le Guern A, Savin C, Pizarro-Cerdá J (2019) Yersinia pestis and plague: an updated view on evolution, virulence determinants, immune subversion, vaccination, and diagnostics. Genes Immun 20:357–370 [Internet]. Available from: https://pubmed.ncbi.nlm.nih.gov/30940874/. Cited 7 Sep 2021
60. Manes C, Gollakner R, Capua I (2020). Could Mustelids spur COVID-19 into a panzootic? Vet Ital [Internet]. Available from: http://www.ncbi.nlm.nih.gov/pubmed/32909703. Cited 17 Dec 2020
61. Pilo P, Frey J (2018) Pathogenicity, population genetics and dissemination of *Bacillus anthracis*. Infect Genet Evol 64:115–125 [Internet]. Available from: https://pubmed.ncbi.nlm.nih.gov/29935338/. Cited 30 Nov 2022
62. Cano-Jiménez E, Acuña A, Botana MI, Hermida T, González MG, Leiro V, et al (2016) Farmer's lung disease. A review. Arch Bronconeumol 52:321–328 [Internet]. Available from: https://pubmed.ncbi.nlm.nih.gov/26874898/. Cited 24 Oct 2022
63. Pereira E, Araujo I, Felipe L, Batista M, Júnior S, Barboza E, et al (2023) RFID technology for animal tracking: a survey. IEEE J Radio Freq Identif 1
64. Ienca M, Scheibner J (2020) What is neurohacking? Defining the conceptual, ethical and legal boundaries. Dev Neuroethics Bioeth
65. Sato H, Maharbiz MM (2010) Recent developments in the remote radio control of insect flight. Front Neurosci 4:199
66. Gkantouna VA, Kambouris ME, Viennas ES, Ioannou Z-M, Paraskevas M, Lagoumintzis G et al (2015) Introducing dAUTObase: a first step towards the global scale geoepidemiology of autoimmune syndromes and diseases. Bioinformatics 31:581–586
67. Kull S, Ramsay C, Lewis E (2003) Misperceptions, the media, and the Iraq war. Polit Q 118:569–598 [Internet]. Available from: http://www.jstor.org/stable/30035697
68. Oliveira M, Mason-Buck G, Ballard D, Branicki W, Amorim A (2020) Biowarfare, bioterrorism and biocrime: a historical overview on microbial harmful applications. Forensic Sci Int 314:110366

69. Tucker J, Sands A (1999) An unlikely threat. Bull Atom Sci 55:46–52
70. Gottschalk R, Preiser W (2005) Bioterrorism: is it a real threat? Med Microbiol Immunol 194:109–114
71. Cole LA (2016) Open-air biowarfare testing and the evolution of values. Heal Secur 14:315–322 [Internet]. Mary Ann Liebert Inc. Available from: http://www.liebertpub.com/doi/10.1089/hs.2016.0040. Cited 3 Feb 2021
72. Clarke SC (2002) Bioterrorism: an overview. Br J Biomed Sci 59:232–234
73. Takahashi H, Keim P, Kaufmann AF, Keys C, Smith KL, Taniguchi K, et al (2004) *Bacillus anthracis* incident, Kameido, Tokyo, 1993. Emerg Infect Dis 10:117–120 [Internet]. Centers for Disease Control and Prevention (CDC). Available from: https://pubmed.ncbi.nlm.nih.gov/15112666/. Cited 10 Dec 2020
74. Carus WS (2002) Bioterrorism and biocrimes: the illicit use of biological agents since 1900. Fredonia Books, Amsterdam
75. Carus WS (2015) The history of biological weapons use: what we know and what we don't. Heal Secur 13:219–255 [Internet]. Mary Ann Liebert Inc. Available from: https://pubmed.ncbi.nlm.nih.gov/26221997/. Cited 30 Sep 2020
76. Schutzer SE, Budowle B, Atlas RM (2005) Biocrimes, microbial forensics, and the physician. PLoS Med 2:e337 [Internet]. Public Library of Science. Available from: https://dx.plos.org/10.1371/journal.pmed.0020337. Cited 10 Dec 2020
77. Jansen HJ, Breeveld FJ, Stijnis C, Grobusch MP (2014) Biological warfare, bioterrorism, and biocrime. Clin Microbiol Infect 20:488–496
78. Schneider DK (2016) China's new legalism. Natl Interest 19–25 [Internet]. Center for the National Interest. Available from: https://www.jstor.org/stable/26557304
79. Leng A, Sun Y (2024) The impact mechanism and breakthrough path of COVID-19 on enterprise financial distress: evidence from China. Econ Anal Policy 82:16–31 [Internet]. Available from: https://www.sciencedirect.com/science/article/pii/S0313592624000572
80. Habibi Z, Habibi H, Mohammadi MA (2022) The potential impact of COVID-19 on the Chinese GDP, trade, and economy. Economies:73
81. Saltzman IZ (2023) Diversionary words: trump, China and the COVID-19 pandemic. Chin Polit Sci Rev 1–25
82. Cooper P, Lampropoulou S (2024) China Virus, Kung Flu, and MAGA: countervalues and sociological fractionation on Twitter as evidenced by pro- and anti-Trump discourses in relation to COVID-19. Discourse Context Media 57:100758 [Internet]. Available from: https://www.sciencedirect.com/science/article/pii/S2211695824000047
83. Evans D, Peattie M (1997) Kaigun: strategy, tactics, and technology in the imperial Japanese Navy, 1887–1941, 1st edn. Naval Institute Press
84. Su Z, Cheshmehzangi A, McDonnell D, Ahmad J, Šegalo S, Xiang Y-T et al (2022) The advantages of the zero-COVID-19 strategy. Int J Environ Res Public Health 8767
85. Han Y, Xu T (2024) Chinese model in global COVID-19 prevention and control: successful experience, institutional advantages and optimal path. Heliyon 10:e25701 [Internet]. Available from: https://www.sciencedirect.com/science/article/pii/S2405844024017328
86. Wu Y, Cao Z, Yang J, Bi X, Xiong W, Feng X et al (2024) Innovative public strategies in response to COVID-19: a review of practices from China. Health care Sci 3:383–408
87. Sturkie T (2023) Must China pay? How claims against China for COVID-19 reveal flaws in the international legal system that make accountability impractical. Penn ST JL INT'L AFF 11:218–220
88. Nakayama Don K (2023) The opium wars of China in the nineteenth century and America in the twenty-first. Am Surg 90:327–331 [Internet]. SAGE Publications. Available from: https://doi.org/10.1177/00031348231192048

Part III
Disruptive Biosecurity

Chapter 6
Understanding and Anticipating the Future (Bio)Threat: The Paradigm Change

Abstract The relative absence of high-profile perpetrated bioevents is a poor predictor for the nature and imminence of the threat. Projections and surveillance based on the Intent of possible perpetrators are easily applicable, but unreliable. Minimal perpetration approaches combined with 5G-Internet assisted preparations and within-establishment operatives devalue the Intent predictor. The identification of malefactors prone to facilitate or commit such acts may be achieved by focusing not on intentions but on capabilities, a more objective predictor. Given the nature of the projected biothreats and their causality, focusing on the weak or choke points of the threat pipeline of a perpetrated event might be advantageous compared to epidemiological projections. This focus on objectively identified and assessed parameters may be achievable by AI. Wide and Narrow AI would assist in tackling Advanced biothreats in terms of risk analysis and in the planning and implementation of responses. The design of diagnostic tests and the selection of the most promising effect (thermotherapies, electrotherapies, radiotherapies) and delivery technologies (inhalable agents, bacterial/viral platforms, biohybrid microrobots and diadermal transfer by patch microneedles) and conventional, injectable and ingestible drugs, for symptomatic and causative therapy, prevention and prophylaxis shall be instrumental in an environment of unforeseen biothreats.

Keywords 5G internet · Self-CRISPRing · Biohacking · Kitchen microbiology · *CrowdStrike* · Improvised contaminating/dispersing devices (ICD/IDD) · Criminogenic potential · Reactive biosecurity · Proactive biosecurity

A focal point in Biosecurity, as with all security and warfare applications, revolves around knowledge, cognition, understanding and, ultimately, intelligence. The usual approach is to survey the unknown, identify possible hostiles, fix them and track

---

This chapter is in concept based on the paper Disruptive biosecurity: A new concept for preparedness and response to the emerging total biothreat landscape by Kambouris ME et al. (2025) published in Comparative Strategy 1–33, https://doi.org/10.1080/01495933.2025.2565586, with updates, corrections, and considerable additions and deletions regarding the original content.

M. E. Kambouris, *Preparing for the Meta-Biothreat: Disruptive Biosecurity*, Advanced Sciences and Technologies for Security Applications,
https://doi.org/10.1007/978-3-032-25690-4_6

their movements. Thus it is the hostility factor that is used to prioritize intelligence missions, from which it is expected to deduce what is planned by such entities, when it is to occur, by whom it is expected to be implemented and with what means. The last is very important, as it defines to a great extend the nature and means of response, of deterrence and of protection. There is a selection of conventional means, and then there is the realm of unconventional means, including, but not limited to, Agents (rather than the much coveted "Weapons") of Mass Destruction, which comprise part of the asymmetrical threat, the ultimate deterrence tool since the invention of the atomic weapons.

As obvious, when the Intent of some entity is considered through some intelligence product to fall in the category of hostility or enmity, resources are assigned by priority; once the capabilities of the hostile are revealed, the response/preparedness/deterrence is formulated accordingly. It obviously does not make sense to treat Intent and Capability as essentially distinct, or rather alternative indicators, but rather as complementary. Hostility with no credible capabilities remains an empty threat and no more resources are to be spent onto it. The Capability without malicious Intent cannot lead to countermeasures, as this creates lack of focus and may refer to friendly or neutral entities. The above might be theoretically valid assumptions, but constitute a poor policy, especially in the case of Biosecurity. The two-stage approach, first to establish Intent and then monitor the Capability, is not sufficient, and it is bound to be even less so in the near future, probably following a geometric regression as Intent may be hidden under smiles and docile behavior, whereas Capability is hard facts and thus its use qualifies as an evidence-based approach, which may be sorely needed already.

6.1 The *Status Quo Ante*

At the end of the first quarter of the twenty-first century, the consensus in biosecurity appears to be rather reassuring. The Security Community seems to insist, since 1999, in classifying the risk for perpetrated bioevents, and their expected threat level, as Low [1–4]. This might refer not to the adjective as such, but to some threat/risk classification system. Such systems usually contain an odd number of levels. A five-level system seems to be a decent compromise between simplicity and accuracy. Despite some controversy over the matter of the number of levels, the use of the word "Low" probably means a "Level 1 out of 5".

At the same time, the Public Health community seems to expect an—unknown—next pandemic, called "Disease-X", probably a spontaneous event, with consensus leaning towards viral origin [5–8]. Such a spontaneous event is considered much more imminent a threat than any perpetrated bioevent. In the wake of the COVID-19 this seems logical, and thus precautionary and preparatory measures for such eventualities, under the moderately sounding caption PHEIC-Public Health Emergency of International Concern, are prioritized over response to perpetrated biosecurity events, which were the centerpiece—although not the sum—of the much more menacingly sounding GCBRs-Global Catastrophic BioRisks of the late 2020s [9–12]).

In all justice, both the above conclusions—or rather hypotheses—seem logical and solidly substantiated. But there are some pitfalls in both. Regarding the waves of spontaneous bioevents of considerable magnitude, a string of epidemics did occur in the twenty-first century like the Ebola, the Avian Influenza and SARS-CoV-1, the SARS-CoV-2 being the climax [5, 13–15]. The direct and indirect impact on individuals and polities has been extraordinary; but not unprecedented. The HIV ravages the planet since early 80s uncontested. Effective treatment came in rather late and refers to the suppression of the progress of the symptoms, not to the clearance of the infection and definitive therapy [16, 17].

In contrast, the few such events of proven human agency caused rather low body-counts [1]. The Aum-Shinri-Kyo Anthrax attack in the 90s was an utter failure [18, 19]; the 2001 Anthrax letters in the US caused single-digit of deaths [20], the 1984 Dalles poisoning by *Salmonella enterica* in no deaths and low tens of hospitalizations [21, 22]. More recently some thwarted attempts in European soil were to be solitary, occasionally lone-wolf operations of low or very low scale with less than a hundred possible deaths—mere "events" by some definition [3]. As a result, natural occurrence seems indeed to hold the highest risk as even the most recent threats had been few, isolated, mismanaged and of small-scale, low-incidence engineered by domestic or immigrant dissidents of known intent, loyalties and affiliations.

Bur reality may be vaguer and cause for concern. If factual evidence establishes human agency of an outbreak, this is definitive. In the opposite direction, an event may only allegedly be considered as non-perpetrated as there can be no definitive proof of it, there is only the lack of proof, which might be reassessed as analytical methodology evolves and reexamines old evidence or detects new. As with archaeology and palaeontology, not *finding* does not mean not *existing*. It is not inconceivable, although it remains unproven and unsettling that unseen hands orchestrated one or more of the recent epidemics and pandemics [23–26]. Precedents may have been the Black Death due to the bioassault against the Black Sea city of Caffa by the Tatars [22, 27], and the Smallpox epidemic that ravaged the natives in North America in the eighteenth century due to British machination [28].

Then comes the focal assumption, that bioperpetrators pose a Low risk [3] because would-be ones are not numerous, remain dispersed in non-communicating nuclei for security purposes, and lack skills due to background, social, and cultural. This rationale is inherently flawed. It associates the status of such risk (supposedly of low priority, despite the sagacity to deliberate on the matter by inter-disciplinary bodies of experts) with own conclusions of the recent past, whence skilled groups with decent engineering and scientific constituents and capable leadership have not been encountered for the time being in perpetrating bioassaults.

But lack of own experience and objective proof does not substantiate lack of existence. It might be a testimony to their skill and abilities, and to the fact that for whatever reason they remain uncommitted regarding relevant assaults. They keep their potential, and build an element of surprise, for the selected opportunity or necessity. Under this assumption, when they are triggered the usual Red Flags (identifiable surveillance markers of imminent danger) might be inadequate or irrelevant. The lack of proof for the existence of such groups, and any indication of their conditional number and numerical strength is not to be taken as reassuring, but as ambiguous when determining the threat level. Consequently, the current deliberation that it is Low might be erroneously reassuring and utterly unproven. A false sense of security and indices of normality constitute reoccurring intentional patterns to throw targeted entities and their specific and collective intelligence off guard, and thus achieve surprise before action is taken. The very different in methods and aims Israeli operation Rising Lion (briefly discussed later in this chapter), employed shooting, transmitting and exploding means of different categories but it remains exemplary to how misplaced is the feeling of security when based on previous experience and not objective predictors.

6.2 Intentionally Culpable

As already mentioned, there are two basic unknown factors in the covertly perpetrated events. *Intent* is the first one and it refers to the motivation of prospective perpetrators and thus, indirectly, exposing their identity, both having been revealed in the abovementioned cases. *Factuality* is the second one and scrutinizes signs of step(s) of the malicious processes before they mature into definitive act. These two factors are both attuned to the current practice: initially focusing on the hostile or criminal *Intention*, so as to detect and identify prospective perpetrators within a naïve population, supportive, tolerant or indifferent. Through such tracks, obtainable through targeted surveillance (the term "Surveillance" is taken in its original context as "setting a factual horizon to apperceive respective new incidents" and not as an epidemiological term), proactive apprehension is possible, before an attempt is ever made. But also of importance is the resulting deterrence to ever contemplate such proceedings. Both the above are parts of the Proactive Biosecurity. Put into a nutshell, surveillance allows detection of hostile/malicious entities first by their Intent, and then follows up by monitoring and assessing their Capability, making Intent the *primary* indicator as things now stand; by no means the *only*. If Intent ticks, then respective Capability becomes of interest as a possible risk factor and processed accordingly.

A dissimilar example, regarding nuclear capability, is tell-tale: the Intent of a state pursuing nuclear technologies (NOT necessarily nuclear weapons, as the latter are banned by a controversial[1] but active treaty[2], whereas the former is a legitimate pursuit based on a fundamental right) changes the-subjective—response and position of other states towards such pursuits. The alleged nuclear arsenal of Israel [29] is

[1] The signatories with nuclear weapons deployed before the conventional date set by the treaty were obliged to gradually delete their own arsenals, none of which ever did so.

[2] The Treaty on the Non-Proliferation of Nuclear Weapons (NPT) of 1970, re-ratified and extended indefinitely since 1995.

a perfect example, but the (intentional) ambiguity on its existence saps the force of the argument. A better example is Japan's enrichment plans, which, contrary to North Korea's, are met with limited international consternation, and this mostly from China. Still, the threat perception lies, objectively, with the Capability. Japanese enrichment [30] may lead to nuclear weapons similarly to what happened with N Korean such endeavors [31], meaning more nations prone to launch nuclear disaster for some pretext or other to some other. But Intent is the regulator: the Japanese are less prone to do so than the N Koreans, and for perfect reasons. And it is also a matter of perspective. The Japanese are less likely to unleash such weapons on the US or the US allies (or subjects, as the Athenian Alliance has resolved such issues millennia ago).

Subjectivity is shown by the comparison between Japan and N Korea to be the utmost liability of the Intent as an indicator. For China, North Korea or Russia (or even South Korea) the picture is different: a nuclear-warfare-capable Japan is considered extremely threatening. N Korea's nuclear arsenal is of no particular concern for China and Russia, and perhaps, in many ways, it is rather advantageous for both.

Still, Subjectivity is not the most serious problem. The fragility of the Intent is much more so, as it is likely to shift literally overnight, (i) with a regime change by a coup, by a decapitation event of different possible causes or by proxy demonstrations, initiated over sanctions and tariffs; or (ii) under a revisionist government. If the capability becomes established, the Intent may follow: Japan is likely at some steep turn of the road of History to stop being so docile as it currently remains towards the US and the Five Eyes [32], since the resentment for more than 80 years of oppression of varying severity might have nurtured silently brewing animosity [33].

Similarly, the Intent may emerge later through violently or not reshaping and rebooting of allegiances, personal or collective. Ukrainian secret services blackmailed civilians, relatives of Russian POWs to perform sabotage in Russia and Russians reportedly do the same [34], although Internet searches reveal only the Russian operations, leaving the spectacularly successful terror and assassination campaigns of the Ukrainian secret services unreported. The capability of both parts to bend the will of said relatives is the key, and the involuntary Russian operatives were chosen by estimates of their Capability in terms of accessing target entities and, bomb, mine, assassinate or otherwise attack them. Thus, Intent of agents is engineered later, *in situ*, while Capability is the defining selection criterion and consequently must be considered as the principal risk factor. A non-state player context may differ in dynamics. Eavesdropping on people and subsequent scrutiny of signs of Capability conditional on apprehension of markers of hostility is more affordable than performing both screenings simultaneously.

The unreliability of the Intent indicator in temporal terms may be seen in the case of Iran in late 70ies: endowed with the most advanced weapon systems vast sums of money could purchase from the US, it became in weeks the bitterest enemy. The efforts of the locally deployed US personnel to deactivate—without reimbursing the owner country/service- were mostly unsuccessful and the repaired and surviving means allowed Iran fighting to a standstill against a well-equipped and widely supported Iraq after 8 years of high-intensity war [35]. One may only wonder what would have happened if Saddam Hussein had opted for an *Entente* with the new, theocratic regime and had attacked Saudi Arabia and Kuwait instead, both at the time sub-standardly equipped. The map of the Middle East would have been different, even if Oman, Yemen and especially Jordan, were not invaded; the latter was prone to direct Israeli land army involvement. The mass of Iranians killed in and by Iraq might have poured in the Arabian Desert fast, using the western automotive equipment acquired by the Shah Pahlavi. Iraq and Syria were thriving and powerful, under secular Baath governances and allied or merely aligned to the still lively USSR and under its protection. The Warsaw Pact at the time was intact, the Soviet invasion of Afghanistan had not yet turned chaotic and the covert destabilization of Poland by the Vatican under the Pole Pope Jean-Paul II and by the Solidarity Syndicate of Lech Valesa were seemingly tackled by the coup of General Jaruzelski in 1981 [36].

Institutionalized enclaves further complicate the *Intent* predictor. Directly organized and committed or outsourced to proxies lies the Corporate Terrorism and Crime [37] or, rather, the "*Within-Establishment* terrorism and crime", whence cardinal assets of the perpetrators' are not dissident or alien, but fully incorporated within the targeted sociopolitical and socioeconomic establishment, if not distinguished members of its elite. These may be decision-makers and other links of the chain-of-command of any level and authority. Their optimal if not privileged positioning differentiates them from lowly recruits such as domestic dissidents or sleeper agents planted beforehand in an environment of prospective or actual interest, as are the operatives recruited *ad-hoc* by Ukraine for having Russian passports, citizenship or mere home address and occupation, who are blackmailed or hired to commit acts of terrorism/hybrid warfare [38].

The Corporate (bio)terrorism is the main constituent of the Bioterrorism iteration of "*Within-Establishment* terrorism and crime". It consists a possible risk of various forms, multiplied by the risk of spontaneous or perpetrated accidents [39]. Cyber entanglement in causality and leverage of impact also increases continuously, as high-containment labs (for basic research, or for mass production) become part of the equation [40, 41]. They should be considered as potential risk foci and are domestic, already dispersed and well-positioned for surging or sequential multi-focus events. In such cases updates in inherent risk level are indispensable, but perhaps of little validity: corporate secrets are mostly kept off range from the regulatory and security inspections meant to assess actual risk levels and the extent of the corporate endeavors [42].

It is much more than simple market manipulation for profiteering or "creation of demand for supplying" proprietary health-related products [43]. Divergent individuals and dissident groups may be found employed within corporate entities of biosecurity interest [44], or such entities may be operated, wholly or partially, as proxies of state and non-state actors set to engineer their decisions or possess their secrets [45, 46]. Space-X offered, for example, free Starlink® connectivity to the Ukrainian military for quite some time after the Russian intervention of 2022. Such generosity was "encouraged" if not mandated by the government which saw in this fact advancements of its aims [47, 48].

Four components are required for a perpetrated bioevent: the bioagent, its delivery mechanism, its target and the perpetrator. From two to all four of them may be found within the same corporate entity. This is only recently, and reluctantly, understood to be a systemic biothreat [49]. The understanding that the Academia fares similarly is even more reluctant, despite it being the second most prominent constituent of Within-Establishment-lurking prospective perpetration. In such environments, any malevolent controller—recruiter is in constant contact with a great number of gullible young persons [50] of variable background and diversified skillsets, mostly state-of-the-art and highly honed ones (both objectively and relatively) [51]. Such institutions, from research centers to university faculties, include some number of accomplished scientists and an even larger of trainees, and in a higher ratio compared to the sum of the enrolled workforce than in the most high-tech corporation.

A further factor would be that oftenly Academia is very critical if not outright hostile to any notion of security, leading to security breaches. The paradigm in this aspect is the famous- or infamous- *Cambridge Five* after World War II, a legendary within-establishment spy ring to date, a blatant security failure within the country with the highest acceptance and performance of security and (counter)intelligence services [52]. The blame game is irrelevant for a failure of such nature and magnitude. What is relevant is that to select, recruit, indoctrinate, train and control flawlessly five brilliant individuals located in a highly prestigious academic institution almost simultaneously, points to one or more proficient handlers(s), who set and ran this operation with said assets.

A repetition is foreseeable as conditions are more favorable: the social dissent is rising and cultural, national and fractional or tribal antagonism if not outright hostility is raging in multicultural and multinational, inclusive campuses and inflamed by competitive logic and blatant inequalities; more so in absence of universally accepted morals and customs. Said conditions facilitate manipulators, indoctrinators and recruiters to achieve considerable penetration and thus leverage for a number of attempts and targets: from regime change as in the Arab Spring [53] and the Color Revolutions [54] (irrespective of success) down to fits of rage and violence by religious or secular basis, as the alleged anti-semitism in US Academia in 2024–2025 [55]. If the leading polity example is interventionist, with invasions, kidnapping and bombing campaigns and piratical practices (or those of corsairs/buccaneers, which is a better legal precedent) violence is understood as the norm and in a proportional function: more is better and there is no such thing as too much.

6.3 The Cradle of Causality: New Iterations of Capability

Recasting biosurveillance might be required. Identifying risk carriers through their ill will and propensity to commit acts of aggression and assault is reasonable and practical to implement proactive biosecurity, given that *Intent* is approachable by standard human surveillance techniques like audio eavesdropping, video recording, wiretapping telecommunications of voice/data (including Internet); and also by human intelligence (HUMINT), such as direct and indirect interrogation. Secret and security services and the Military, have accumulated centuries' worth of accumulated experience of the kind and applying such methods is relatively commonplace, uncomplicated, standard and affordable, with no extra risks or unknowns. Contrary to that is the need to analyze scientific data and interpret qualitative and quantitative markers so as to properly gauge establishing of Capability.

Neither *Intent* nor *Motive*, a much plainer marker, are reliable proactive identifiers of malefactors. The advent of proxies starting with the Red Terror throughout western countries (with some notable exceptions) in 1970s to 1990s [56], is followed by the rise of ISIL and then ISIS in 2010s [57] and the events in Ukraine and Middle East from the 2000s all the way to 2020s, the two last ones up-scaling to being true wars in duration, intensity, participants and cost, saps the validity of obvious or known intent as a predictor for perpetrated events. The operation Rising Lion against Iran in 2025 is a perfect example [58]. A wave of targeted, coordinated, dispersed attacks by multiple nuclei of Special Forces, dissidents, enemy agents and oppressed minorities using crude and also sophisticated effectors (not only weapons) inserted fully assembled or in parts and assembled locally crippled vital and defended, protected or hidden assets. There was no indicator of such Capability or Intent by the acting agents; nor in the mass murder of Iran's proxies in Lebanon by command-exploding of cell phones and pagers. Dissidents and proxies were known to be active in Iran, but shots, stabs and bombs were the norm in terms of used effectors. Remotely fired guided missiles and even more remotely guided drones launched from relatively close range against air defense assets was not considered within their Capabilities, and was not scrutinized in its specifics. A usual weapon was the massive, violent demonstrations which by 2026 was upscaled to effect regime change by the obvious controllers of such proxies.

The complications brought about by proxies in conventional or special and hybrid operations apply to the full range of bioincidents, from Biocrime, through Bioterrorism, to Biowarfare [59–62], as the respective perpetrators may remain under the Intent-informed awareness horizon. As with Iran, perpetrators of skill would keep a low profile or adopt misleading behavior, the paradigm being the false religious and social patterns advised by the Muslim *taqiyah* and the Jewish *marrano/anusim* for surviving the Inquisition. With Intent becoming unapproachable, Capability constitutes the only usable predictor. It follows the rationale behind the suggested objective reconsideration of the biothreats, by measurable qualities of bioagents [63] and is conceptually similar with established and successful security practices for air travel. The knowledge of the minds of the passengers are unapproachable to the airport security. Still, the means to carry out illicit or criminal aspirations must be carried on their persons or in their baggage and testing and scanning such entities should do the trick. Material signatures of interest are olfactory and magnetic ones, associated with narcotics, explosives and metal respectively; while form factors suiting bladed instruments and firearms are prime indicators. The most malicious intent cannot materialize without proper means. The idea remains valid despite many more objects being prone to be used, from hand-made unconventional explosives to highly concentrated acids and corrosives or poisons and pathogen preparations. Threat factor increases not by the will and purpose to acquire and use destructive means (Intent) but by the access secured to said means regarding hardware and know-how.

The new vogue in biosecurity awareness could summarily reject background and deliberately introduced noise sources, like agent selection criteria, as the prospective perpetrators may have widely different priority lists, and these would be unknown to the responders and most probably would consist objects of guesswork, educated or not, and prone to deliberate misinformation. The constant lies to the sheer necessity that as with any weapon, the bioagent follows the binary scheme vector-effector, both parts of which may be multi-stage and of dissimilar nature. The various bacteria-based delivery systems of cytotoxic compounds to tumors [64–67] exemplify said dissimilarity. Translated into criminal intent and application, the sheer number and variety of potential end effectors, meaning the active compounds to be delivered to different biocompartments, instead of tumors, to cause septic or toxic effects, if combined with the abovementioned bacterial vectors, is exorbitant. Impossible to guess the combination a secretive and tight-lipped perpetrator would opt for, even if using AI to make the guess. Moreover, the said bacterial vectors must be inserted in the target organisms somehow. This implies multiple vectoring stages, like aerosols released by some pneumatic or explosive devices, or spiked cutting/hollow piercing instruments.

The focal issue is the need of the aforementioned *pair*, as there is an almost indefinite selection of effector agents, created by the continuous developments in the fields of Genetic Engineering [68], Synthetic Biology [69] and Xenobiology [70]. As a result, to somehow foretell the identity of the selected agent, even roughly, and thus concentrate efforts in a proactive manner against a series of bioagents prioritized by probability, is conditionally almost impossible—although for more limited threat scenarios it may be workable due to sets of restrictions faced by different *known* perpetrators. But in principle, foretelling the agent of choice and in detail enough to produce proactively existing or novel diagnostic amenities ("novel" not meaning of new technology, simply of optimized design and performance) and suggest therapeutic or protective regimens beforehand is out of the question. The development of *de novo* therapies is time-ineffective as seen with SARS-CoV-2, vaccines included [71, 72] and retasking in a reactive manner seems a fair option, although, once more, conditionally so.

Similar diversity and range of options is seen in selecting vectors. Live ones include hosts of different types (like Definitive, Intermediate, Main, Secondary, Paratenic, Accidental) [73]. The variety of technical delivery means, in essence fomites but also dispersion systems, is even wider: intercontinental carriers, mostly missiles; space-based dispersion options (the privatization of space launching private loads into orbit affords ground for extremely unpleasant surprises), conventional systems (ammunition and platforms) with suitable provisions or modifications (aircraft bombs, artillery and mortar shells, rockets, spraying aircraft and drones) easily come to mind. But there are also the local delivery amenities: fragmentation devices, bullets and thrust/cut weapons tainted with bioactive preparations of different kinds (toxic-infective) and other, improvised multi-dispersed delivery approaches, possibly but not necessarily implemented by multiple human agents, cognitively or not.

Multiple human agents are the lowest tech possible and may be approached remotely and, before or after indoctrination and conditional to its results, recruited. Direct such proceedings (but indirect and remote as well) are applicable for the masses of immigrants, although already recruited and trained individuals might have been inserted among the anonymous mass beforehand and at ease. Any numbers approached by these two pipelines may be equipped with house or gardening amenities, like sprayers of different kinds and capacity (house-cleaning hand-portable ones, gardening sprayers on pick-up trucks). They may also improvise (by using Dark Internet encrypted and mailed or hand-delivered printouts of manuals) timer—or remote-controlled Improvised Dispersion/Contamination Devices (IDD/ICD, as equivalents to IED—Improvised Explosive Devices) [58]. For utmost simplicity and really substandard operatives, unexposed to Internet Warding by Security and Law Enforcement agencies, the use of spiked edged weapons or utensils and equipment would be a fair option. All the above are bound to provoke multiple focuses of disease, epidemic or not. It is of cardinal importance to brief trauma care facilities and personnel, especially when attending surging or massive terrorist-incurred wounds that these may be intentionally infected (and most probably by something more dangerous than the usual choice, a bioregulator or, more often, a toxin, like ricin), [74] and thus physicians and nurses must operate under specific protocols when offering wound care.

One such operative may cause several tainted injuries if operating within crowds as in public transportation or sports, shows, political gatherings and other events of free and massive attendance. The terror and the death and injury toll would be significant, but even more so the delayed effect when non-fatal injuries are treated and the victims are at home, in diverse locations within a city or even state. They may start presenting symptoms and signs, and transmitting the disease if the delivered pathogens are communicative and have a quiescent stage long enough. The initial delivery mechanism from the fomite-weapon to the original victims would be some kind of fast-drying dye with spore or other particle (i.e. virion or even vegetative cell) suspension applied to the edges. The contact with body fluids and their pH or other defining factor would liquefy the dye and release the spores/cells/particles into the wound. The transformation from quiescent to vegetative status would increase the delay, possibly until their return to home and work, where they may start transmitting the infectious agent, causing a multi-focus event; especially lightly wounded individuals in massive events are expected to be tended summarily and released soon from the attending hospitals and First Aids facilities to make room for more.

The underlying technology of the above engineering is not particularly exotic. If firearms are to be used, the dyed preparation must be thermotolerant; a version of the technology used in the BioBullets that are employed to vaccinate wildlife-especially fast or dangerous species—from afar [75]. For contact weapons such properties are not needed. Moreover, the firearms may be deformed, to make identification by standard surveillance and scanning more difficult. This is much easier when they are produced by 3-D printing, a much safer approach than smuggling, and infinitely more flexible, not requiring the skill of gunsmiths [76]. Still, smuggled firearms are always an option, especially in unguarded frontiers and corrupted polities. Hand guns are small and light, easily concealable. Some more powerful weapons, like submachine guns are similarly formed: they are compact and light, marginally less so than handguns, but sport high rates of fire, capacious magazines for sustained fire and if combined with dispersible ammunition, to the like of shot in the (much more abundant but cumbersome to conceal and use) shotguns, the number of wound-infected casualties may become quite massive by a small number of perpetrators operating randomly or in a coordinated manner, in surging or dispersed format.

In general and without quantitative and qualitative criteria the Capability for perpetrated bioevents is actually widespread. Secluded domains are of extreme interest as mentioned above with respect to the Industry and the Academia. Still, in many cases such Capability has proliferated considerably: the *Black Biology* is the paradigm and implies biology used for any kind of malicious purpose, "malicious" taken in any applicable definition [77, 78]. *Biohacking* is another case, a current term insinuating any unauthorized, spontaneous such efforts but incorporating not only actual, wet procedures but also bioinformatics, used in conjuction with wet biochemistry or alone as detailed later and in Chap. 5 [79]. The older concept of *kitchen microbiology* is included in the former trait of Biohacking and is self-explained. It refers to mostly unmodified microbiota and substitutes expertise of the implicated operatives for the scarcity of equipment and instrumentation for Conventional Microbiology purposes, such as culturing of microbiota and their subsequent purification.

More recent amongst the Biohacking branches is the *do-it-yourself-biology* [80], which is the elementary level as it presupposes only one, perhaps expert, operative, operating in solitude and thus very difficult to uncover by the standard interruption technique of gradual infiltration. The self-CRISPRing is a current trait. It aims to *genetically modify* naturally occurring organisms or *enhance cyborgs* and *fully engineered bioorganisms*. Both aims are tricky due to the practice of widespread physical and virtual surveillance, especially of purchases. Some required supplies, like the enzymes, are not sold openly, as the suppliers self-regulate their merchandise and are conscious of the payment methods and dispatch addresses. Any effort to bypass such difficulties leaves paper or digital trail, which may be picked by proper surveillance setups. The best example is on-the-spot cash payments, which are always suspect and can be processed so as to initiate security cameras in the cashier's office; the taken video or frames would be subject to automatic, real-time, or subsequent, on-request correlation with any other clue or cue (mainly by comparing different records), to identify a face and attribute some cumulative risk score.

6.4 The Virtual Amplifier of the Perpetrated Biothreat

The proactive pathway in Biosecurity might be inapplicable in the emerging context. Operating based on assumptions and risk levels coming out of them, it faces reality checks revealing liabilities. The abovementioned supposed lack of high-end skills by the prospective bioperpetrators, falsely depreciates the actual threat, despite all Intent indicators, which in today's surveillance toolkit, stand of cardinal importance. This erratic assessment is based on some assumptions. One is the need of competences and knowledge acquired only by advanced, extensive, even if crash-course training. Knowledge is needed for conception and planning some endeavor, competence is needed to carry it out. Still these two functions and thus their objects may be decoupled: some may plan and conceive and thus need knowledge, and others may engineer, needing skills and practical knowledge. Another assumption is that the multi-level footprint inherent in staging such events would be picked by routine surveillance assets. For example, in the 60s and 70s prospective terrorists used to visit international camps in supportive/sponsoring or failed states to undergo training and indoctrination, which provided further motivation and confidence in the skills they were developing [81]. Interested parties could track, follow and map such movements. Modern-day practice informs against such excursions, through high-tech, intrusive surveillance and denial of free travel. The last is implemented by targeted arrests, confiscating of passports, occasionally on weak pretext and without due process, and also by the constant upsetting of travel planning by sudden bans of specific connecting flights for any weak or strong security prerogative. Similarly, the curtailing of money laundering and the constant surveillance of transactions make the financing of such endeavors problematic and sparse. The solution has come from the 5G Internet that makes such training trips redundant, as detailed in Chap. 5, especially if combined with different applications of Artificial Intelligence-AI. Thus, out-of-the-establishment prospective perpetrators, being non-integrated and hostile aliens or disenfranchised dissidents, dwelling within the target state/community, may be recruited and trained mentally and functionally from afar, as already mentioned in Chap. 5 and assisted considerably in the field of applied logistics.

The unimpressive skillset of apprehended current-generation aspiring perpetrators is circumstantial: they are mostly fanatics whose zeal is not matched by their competences. This may be intentional misleading and noise to derail surveillance and response algorithms. Competent agents, in cells, or even operating as lone wolves, would be better prepared and outfitted, competently guided and perhaps in-position already. The ineptitude of terrorists allows a war footing of societies but also permits the globally interdependent markets and associate societies to feel secure and safe in order to operate as usual, without panic-driven disruption. High-profile and high-profit events, like championships, games, concerts and Markets' daily routine plus transportation, trade and travel are vital not to be interrupted or degraded by fear, as they are existential needs for the running and survival through supply of the modern Metropoleis of tens of millions.

An additional issue pertaining the virtual amplification is the minimal or discrete perpetration, i.e. the use of technidal leverages to orchestrate a bioevent much more efficiently in terms of engagement and resources than in times past. Less perpetration is needed for a bioevent of a given magnitude [82], as detailed in Chap. 5. This liability shall progress by the existential role of the 5G Internet in daily life and all evolution or perversion set to occur in societies and daily lives by its agency. Almost every aspect of activity will be implemented remotely, in the beginning to assist motion-challenged people, and then mandatorily for ecological reasons, or without any reasoning as with the dematerialization in transactions and the use of Credit instead of Money. Three enabling developments were required for such proceedings: the Internet-of-Things, extreme automation and Big Data handling—the very concept of Industry 4.0 [83]; the Artificial Intelligence was not needed in principle, but shall magnify their individual and combined impact.

Minimally perpetrated bioevents include the perpetrated accidents whence release of a pathogen might be effected by an intentional degradation of safety mechanisms and procedures or a direct sabotage, as an explosion in a containment area while work proceeds with transmittable and virulent agents. They do not need to be the result of cyberoperations that may bring about safety failures. But this particular breed of perpetrated bioaccidents consists the cyber-enabled (or cyber-perpetrated) bioevents mentioned in Chap. 5. It is a highly sustainable approach whence efforts to select, manipulate/process and finally deploy a bioagent are redundant. A suitable bioagent might be available and perhaps located advantageously to cause a disruptive and possibly a destructive effect, had it not be for the containment and security/safety measures preventing its escape and transmission by fomites and/or vectors. If such events may be produced by cyberoperations rather than Special Forces infiltrating and blowing up or sabotaging containment features, the risk relates to (one definition of) cyberbiosafety [84], irrespective of any use of AI, and may result in a true bioevent.

Then comes the cyber-leveraged bioevent. A given bioevent, its origin being inconsequential, is magnified in perception and impact by actions undertaken in the cybersphere; this (virtual) location of the causality and/or of the magnitude allows inclusion within the concept of *infodemics* [85]. The magnitude and alleged impact of the original bioevent is of some consequence, as its character lends it more or less to magnification. The latter is achieved by engineering failure of intervention, especially treatment (utter, through suppression and denial; or partial, through misapplication). The pathway to implement such effects in various degrees is to negate or falsify communications and command assets and traffic of any kind. The current iteration would be through AI as discussed in Chap. 5. Still, plainer iterations exist: to undermine and degrade Information Technology (IT) services applicable in biosurveillance, agent detection and identification, and spatiotemporal delineation of an outbreak; in deciding upon, distributing and prescribing prevention/treatment, providing counselling, and any kind of support to affected natural individuals and legal entities [86, 87]. The IT blackout of 7/2025 due to the Windows®/*CrowdStrike*®, although it had preciously little to do with health support, biosecurity and biosafety, allowed a peek in the brave networked world [58]. Jamming of communications, an all-times favorite, may still degrade response, though not as fully as the denial of IT services, and actual, not virtual, sabotage in ambulances, hospital cablage and corruption of medical supplies by various means, all promote to different degrees the abovementioned objective. In the individual and communal level, bionegationism [88], biodenialism [89, 90] and misinformation [61] could upset response and add noise to surveillance, the example being clinical and spatiotemporal misreporting of cases that may compromise syndromic surveillance needed to inform response.

The *agentless* (bio)event is a case of infodemic and in fact it is not a bioevent as there is no bioagent involved; still, secondary destructive results and disruptive ones may be similar to the ones produced by its use. The idea is how to effect the impact of a bioagent without the expense, trouble and uncertainty of procuring, finding and using one. The less the trouble in the above steps, the higher the unreliability or ineffectiveness of the bioagent; the latter is defined as a bioentity causing not some effect through influence ("priming") but mechanistically and directly so. Effects severely detrimental in nature for a given target set that initiate downstream a cascade of wider events. The usual toolkit of Psychologcal Operations (PsyOps) and disinformation might well achieve the objective; once more, though, AI makes such pursuits easier to achieve. The trick is on one hand to disseminate fake reports by multiple platforms and channels that corroborate each other on the fact there has been a bioevent that incurs considerable casualties and/or material/immaterial/financial damage; and on the other to slander the authorities that they attempt to cover-up the condition and lie to the Public by denying such an event. The combination has the potential to initiate in the real world panic and disaffection/loss of trust, however unjustified, which are prone to amplification in the virtual domain. When panic turns to terror, disruption follows with bank runs, looting of pharmacies and hostility against health professionals for not acknowledging the projected calamity, nor providing assistance. Chaos, conspiracy theories and riots may follow suit as activists and hostile crowds would assault in virtual and real worlds any "appeaser" or "moderator". This is a predictable side-effect in polities where the fear-mongering has been elevated to official policy and repeatedly exited the general public, while opposite views were summarily rejected, suppressed or even prosecuted as *conspiracy theories, disinformation* and *fake news* [26, 61] to promote some or other agenda, including but not limited to instilling fear and create a notion of perpetual threat.

The massive hacking of software-operated devices used for life support and environmental management falls in the healthcare cybersecurity threats [84]; but *sensu lato* they could be considered cyber-engineered bioevents. The possibilities are staggering: deliberate mismanagement of the air-conditioning in a building is likely to cause injuries or sickness by extreme temperatures, or dizziness by not renewing stale air. Bugging pacemakers, defibrillators and other life support equipment may result directly or indirectly in deaths; if health enhancement equipment is targeted, like neurotransplants [91], injuries rather than fatalities should be expected-mostly. And then comes neurohacking [79, 92], i.e. behavior manipulation and forced compliance of carriers of neurotransplants, animals included, to execute orders despite their own will. Orders may involve aggression (targeted or rampant), carriage, or patrolling and intercepting (or reporting). Subjects may include humans [91], stray or owned company animals and wildlife, like sharks, dolphins, birds, insects, wolves and bears [58, 93–98].

6.5 Evolutionary Shift and Paradigm Change

The future perpetrated biothreat might have little in common with the current respective cognitive construct. The current and highly complicated classification by causative agents would be streamlined to a binary pair based on the intended impact (destructive-disruptive) and a three-fold scheme based on the nature of the agency/perpetration. That would be either agentlessly perpetrated; or by the direct or indirect employment of existing virulent agents [61, 86, 87]; or, finally, by the direct use of custom-produced agents, modified or fully engineered may they be [99]. The expected variability of the effectors as described above negates prediction, especially if combined with the multiple vectoring options available through technology and innovation or mere improvisation. Combinations verge on Big Data, contrary to the handful of bioagents considered weaponizable some short decades ago (even in early 2000s), a scarcity used to allow narrowly focused surveillance and diagnostics [63], within the context of Reactive Biosecurity.

Once technology exponentially increased the possible pathogens to be used offensively [100], and this by motivated individual or collective perpetrator entities, biosecurity had to refocus: Away from resilience towards the projected effectors (Reactive Biosecurity), and onto the perpetrators proper. The Intent was a mediocre but applicable determinant for the new iteration, the Proactive Biosecurity.

In Chaps. 3–5 the currently projected threat iteration has been presented: Addition of new threat modalities and kinds and massive expansion of the existing ones. The continuous relevance of previous threat entities is prone to justify inertia in cognitive response to threat status and promote doctrinal and procedural, if not technical, conservatism. As the new threat prerogative is expected to be a decentralized massiveness intended to support intense and sustained but extremely distributed operations, including the use of proxies or alien contractors, lacking discernible Intent determinants, Capability seems better suited to become the threat predictor of choice. It applies both to stern ideologues and faithless mercenaries or even sympathetic proxies in need of cash and patronage, as seen in the Russo-Ukrainian conflict, the example being the extremely brutal and bloody attacks of early 2024 by ISIS-K against targets only marginally suiting its declared agenda.

References

1. Blum M-M, Neumann P (2020) Corona and bioterrorism: how serious is the threat? [Internet]. War rocks. Available from: https://warontherocks.com/2020/06/corona-and-bioterrorism-how-serious-is-the-threat/
2. Tucker J, Sands A (1999) An unlikely threat. Bull Atom Sci 55:46–52
3. Millett P, Snyder-Beattie A (2017) Existential risk and cost-effective biosecurity. Health Secur 15:373–383
4. Gottschalk R, Preiser W (2005) Bioterrorism: is it a real threat? Med Microbiol Immunol 194:109–114

5. Adlhoch C, Alm E, Enkirch T, Lamb F, Melidou A, Willgert K et al (2024) Drivers for a pandemic due to avian influenza and options for one health mitigation measures. EFSA J 22:e8735
6. Lawrence P, Heung M, Nave J, Henkel C, Escudero-Pérez B (2023) The natural virome and pandemic potential: disease X. Curr Opin Virol 63:101377
7. Mitu RA, Islam MR (2024) The current pathogenicity and potential risk evaluation of Marburg virus to cause mysterious "disease X"-an update on recent evidences. Environ Health Insights 18:11786302241235808
8. Yang S, Kar S (2024) Protracted molecular dynamics and secondary structure introspection to identify dual-target inhibitors of Nipah virus exerting approved small molecules repurposing. Sci Rep 14:3696
9. Kambouris ME (2023) Global catastrophic biological risks in the post-COVID-19 world: time to act is now. OMICS 27:153–170 [Internet]. Available from: https://pubmed.ncbi.nlm.nih.gov/36946656/. Cited 24 Jul 2023
10. Schoch-Spana M, Cicero A, Adalja A, Gronvall G, Kirk Sell T, Meyer D et al (2017) Global catastrophic biological risks: toward a working definition. Health Secur 15:323–328
11. Kambouris ME, Manoussopoulos Y, Kantzanou M, Velegraki A, Gaitanis G, Arabatzis M et al (2018) Rebooting bioresilience: a multi-OMICS approach to tackle global catastrophic biological risks and next-generation biothreats. Omi A J Integr Biol 22:35–51
12. Millett P, Snyder-Beattie A (2017) Human agency and global catastrophic biorisks. Health Secur 335–336
13. Morens DM, Fauci AS (2013) Emerging infectious diseases: threats to human health and global stability. PLoS Pathog 9 [Internet]. Public Library of Science. Available from: https://www.ncbi.nlm.nih.gov/pmc/articles/PMC3701702/. Cited 25 Sep 2020
14. Farahat RA, Abdelaal A, Shah J, Ghozy S, Sah R, Bonilla-Aldana DK et al (2022) Monkeypox outbreaks during COVID-19 pandemic: are we looking at an independent phenomenon or an overlapping pandemic? Ann Clin Microbiol Antimicrob 21:1–3 [Internet]. BioMed Central Ltd. Available from: https://ann-clinmicrob.biomedcentral.com/articles/10.1186/s12941-022-00518-2. Cited 1 Dec 2022
15. He J, Kam Y-W (2024) Insights from avian influenza: a review of its multifaceted nature and future pandemic preparedness. Viruses 16
16. Ghosh AK (2023) Four decades of continuing innovations in the development of antiretroviral therapy for HIV/AIDS: progress to date and future challenges. Glob Health Med 194–198
17. Deeks SG, Lewin SR, Havlir DV (2013). The end of AIDS: HIV infection as a chronic disease. Lancet 382:1525–1533
18. Tin D, Sabeti P, Ciottone GR (2022) Bioterrorism: an analysis of biological agents used in terrorist events. Am J Emerg Med 54:117–121
19. Takahashi H, Keim P, Kaufmann AF, Keys C, Smith KL, Taniguchi K et al (2004) *Bacillus anthracis* incident, Kameido, Tokyo, 1993. Emerg Infect Dis 10:117–120 [Internet]. Centers for Disease Control and Prevention (CDC). Available from: https://pubmed.ncbi.nlm.nih.gov/15112666/. Cited 10 Dec 2020
20. Borio L, Frank D, Mani V, Chiriboga C, Pollanen M, Ripple M et al (2001) Death due to bioterrorism-related inhalational anthrax: report of 2 patients. J Am Med Assoc 286:2554–2559 [Internet]. American Medical Association. Available from: https://pubmed.ncbi.nlm.nih.gov/11722269/. Cited 10 Dec 2020
21. Török TJ (1997) A large community outbreak of Salmonellosis caused by intentional contamination of restaurant salad bars. JAMA 278:379 [Internet]. American Medical Association (AMA). Available from: https://pubmed.ncbi.nlm.nih.gov/9244330/. Cited 10 Dec 2020
22. Galatas I (2017) The misuse and malicious uses of the new biotechnologies. Ann des Mines - Réalités Ind. 2017:103–108 [Internet]. Institut Mines-Télécom. Available from: https://www.cairn.info/load_pdf.php?ID_ARTICLE=RINDU1_171_0103
23. Bence Farkas C, Gábor Dudás C, Gergely Csaba Babinszky C, Lászí Földi C (2023) Analysis of the virus SARS-CoV-2 as a potential bioweapon in light of international literature. Mil Med 188:531–540. https://doi.org/10.1093/milmed/usac123

24. Holmes EC, Goldstein SA, Rasmussen AL, Robertson DL, Crits-Christoph A, Wertheim JO et al (2021) The origins of SARS-CoV-2: a critical review. Cell 184:4848–4856 [Internet]. Elsevier. Available from: /pmc/articles/PMC8373617/. Cited 5 Aug 2022
25. Harrison NL, Sachs JD (2022) A call for an independent inquiry into the origin of the SARS-CoV-2 virus. Proc Natl Acad Sci USA [Internet]. Available from: https://www.pnas.org/doi/abs/10.1073/pnas.2202769119. Cited 7 Oct 2022
26. Nattrass N (2023) Promoting conspiracy theory: from AIDS to COVID-19. Glob Public Health 18:2172199 [Internet]. Taylor & Francis. https://doi.org/10.1080/17441692.2023.2172199
27. Wheelis M (2002) Biological warfare at the 1346 siege of Caffa [Internet]. Emerg Infect Dis 971–975. Centers for Disease Control and Prevention (CDC). Available from: https://pubmed.ncbi.nlm.nih.gov/12194776/. Cited 28 Sep 2020
28. Patterson KB, Runge T (2002) Smallpox and the Native American. Am J Med Sci 323:216–222 [Internet]. Lippincott Williams and Wilkins. Available from: http://www.amjmedsci.com/article/S0002962915344815/fulltext. Cited 25 Sep 2020
29. BenLevi R (2022) The evolution and future of Israeli nuclear ambiguity. Nonproliferation Rev 29:243–265 [Internet]. Routledge. https://doi.org/10.1080/10736700.2023.2215583
30. Kim SC (2017) Endangering alliance or risking proliferation? US–Japan and US–Korea nuclear energy cooperation agreements. Pacific Rev 30:692–709 [Internet]. Routledge. https://doi.org/10.1080/09512748.2017.1293715
31. Kim SC, Cohen MD (eds) (2017) North Korea and nuclear weapons: entering the new era of deterrence [Internet]. Georgetown University Press. Available from: http://www.jstor.org/stable/j.ctt1ps318b
32. Gallagher M (2025) Recalibrating the five eyes alliance. Wash Q 48:163–185 [Internet]. Routledge. https://doi.org/10.1080/0163660X.2025.2555719
33. Untalan CY (2024) Beyond empire: Okinawa and the politics of American Military bases in Japan. Polit Anthropol Res Int Soc Sci 5:62–80 [Internet]. Brill, Leiden, The Netherlands. Available from: https://brill.com/view/journals/pari/5/1/article-p62_005.xml
34. Skachko I (2025) They let you talk to a prisoner and then blackmail you into committing sabotage [Internet]. Kharkiv Human Rights Protection Group. Available from: https://khpg.org/en/1608814684. Cited 14 Jan 2026
35. Murray W, Woods KM (2014) The Iran–Iraq war: a military and strategic history [Internet]. Cambridge University Press, Cambridge. Available from: https://www.cambridge.org/core/product/7C6E42D57383472EA9B9F6101BEABD94
36. Eberts MW (1998) The Roman Catholic Church and democracy in Poland. Eur Asia Stud 50:817–842 [Internet]. Taylor & Francis, Ltd., University of Glasgow. Available from: http://www.jstor.org/stable/153894
37. Scott Craig R, Kang Katie K (2023) Shells, fronts, astroturfing, and beyond: examining concealment strategies of proxy organizations. Manag Commun Q 38:623–650 [Internet]. SAGE Publications Inc. https://doi.org/10.1177/08933189231222471
38. Bukhari SRH. Disappearing command: assassinations, purges, and the erosion of Russian military leadership in the Ukraine war. J Soc Sci Rev 5:97–106 [Internet]. https://doi.org/10.62843/jssr.v5i3.566
39. Castro Rodriguez DJ, Shi H, Vitale M, Demichela M, Barresi AA (2025) Assessing the major industrial accident potential based on dangerous substance criteria: a framework for Seveso and Non-seveso facilities. Chem Eng Trans 116:409–414 [Internet]. SE-Research Articles. Available from: https://www.cetjournal.it/index.php/cet/article/view/CET25116069
40. Huffman JA, Perring AE, Savage NJ, Clot B, Crouzy B, Tummon F et al (2020) Real-time sensing of bioaerosols: review and current perspectives. Aerosol Sci Technol 54:465–495 [Internet]. Taylor & Francis. https://doi.org/10.1080/02786826.2019.1664724
41. Alawida M, Omolara AE, Abiodun OI, Al-Rajab M (2022) A deeper look into cybersecurity issues in the wake of COVID-19: a survey. J King Saud Univ Comput Inf Sci 34:8176–8206
42. Morten C (2023) Publicizing corporate secrets. Univ Pa Law Rev 171:1319–1404
43. Wagner AD, Njuguna IN, Neary J, Lawley KA, Louden DKN, Tiwari R et al (2023) Demand creation for HIV testing services: a systematic review and meta-analysis. PLoS Med 20:e1004169

44. Bell AJC, Rogers MB, Pearce JM (2019) The insider threat: Behavioral indicators and factors influencing likelihood of intervention. Int J Crit Infrastruct Prot 24:166–176. [Internet]. Available from: https://www.sciencedirect.com/science/article/pii/S1874548218300982
45. Strauss M (2018) Hostile business and the sovereign state: privatized governance, state security and international law, 1st edn. Routledge
46. Schoch CL, Seifert KA, Huhndorf S, Robert V, Spouge JL, Levesque CA et al (2012) Nuclear ribosomal internal transcribed spacer (ITS) region as a universal DNA barcode marker for fungi. Proc Natl Acad Sci 109:6241–6246. [Internet]. Available from: http://www.pnas.org/cgi/doi/10.1073/pnas.1117018109
47. Sabbagh D (2023) Fury in Ukraine as Elon Musk's SpaceX limits Starlink use for drones [Internet]. Guard. Available from: https://web.archive.org/web/20230309133527/https://www.theguardian.com/world/2023/feb/09/zelenskiy-aide-takes-aim-at-curbs-on-ukraine-use-of-starlink-to-pilot-drones-elon-musk
48. Joscha A (2024) Private infrastructure in geopolitical conflicts: the case of Starlink and the war in Ukraine. Eur J Int Relations 30:842–866 [Internet]. SAGE Publications Ltd. https://doi.org/10.1177/13540661241260653
49. Radosavljevic V (2019) Environmental health and bioterrorism. Encycl Environ Heal 450–457
50. Roberts K (2015) Youth movements (Wright JDBT-IE of the S& BS ed). Elsevier, Oxford, pp 837–841. Available from: https://www.sciencedirect.com/science/article/pii/B9780080970868321742
51. Special Committee on Academic Freedom and National Security in a Time of Crisis (2003) Academic Freedom and National Security in a Time of Crisis [Internet]. Washington DC. Available from: https://www.aaup.org/report/academic-freedom-and-national-security-time-crisis
52. Sinclair A (1987) The red and the blue: intelligence, treason and the universities. Coronet Books, Hodder and Stoughten, London
53. Kumaraswamy PR (2011) The Arab spring. India Int Cent Q 38:52–62 [Internet]. India International Centre. Available from: http://www.jstor.org/stable/44733617
54. Mitchell LA (2012) The end of an era. Color revolutions [Internet]. University of Pennsylvania Press, pp 187–199. Available from: http://www.jstor.org/stable/j.ctt3fj606.11
55. Volodarsky S, Wright G, Hecht S, Saxe L (2025). What crosses the line? Understanding the boundaries of antisemitism and criticism of Israel through the perspectives of Jewish College Students. J Jewish Educ 91:107–131 [Internet]. Routledge. https://doi.org/10.1080/15244113.2025.2476546
56. Geipel GL (2007) Urban terrorists in Continental Europe after 1970: implications for deterrence and defeat of violent nonstate actors. Comp Strateg 26:439–467 [Internet]. Routledge. https://doi.org/10.1080/01495930701750273
57. Fishel KL (2021) The fate of ISIS. Am Intell J 38:74–80 [Internet]. National Military Intelligence Foundation. Available from: https://www.jstor.org/stable/27168701
58. Kambouris ME, Manoussopoulos Y, Velegraki A, Patrinos GP (2026) Disruptive biosecurity: a new concept for preparedness and response to the emerging total biothreat landscape. Comp Strateg 1–33 [Internet]. Routledge. https://doi.org/10.1080/01495933.2025.2565586
59. Oliveira M, Mason-Buck G, Ballard D, Branicki W, Amorim A (2020) Biowarfare, bioterrorism and biocrime: a historical overview on microbial harmful applications. Forensic Sci Int 314:110366
60. Jansen HJ, Breeveld FJ, Stijnis C, Grobusch MP (2014) Biological warfare, bioterrorism, and biocrime. Clin Microbiol Infect 20:488–496
61. Bernard R, Bowsher G, Sullivan R, Gibson-Fall F (2020) Disinformation and epidemics: anticipating the next phase of biowarfare. Health Secur 19 [Internet]. Mary Ann Liebert Inc. Available from: https://pubmed.ncbi.nlm.nih.gov/33090030/. Cited 2 Mar 2021
62. Carus WS (2002) Bioterrorism and biocrimes: the illicit use of biological agents since 1900. Fredonia Books, Amsterdam
63. Colf LA (2016) Preparing for nontraditional biothreats. Health Secur 14:7–12 [Internet]. Available from: https://pubmed.ncbi.nlm.nih.gov/26863312/. Cited 22 Apr 2022

64. Lin D, Feng X, Mai B, Li X, Wang F, Liu J et al (2021) Bacterial-based cancer therapy: an emerging toolbox for targeted drug/gene delivery. Biomaterials 277:121124. [Internet]. Available from: https://www.sciencedirect.com/science/article/pii/S0142961221004816
65. Lou X, Chen Z, He Z, Sun M, Sun J (2021) Bacteria-mediated synergistic cancer therapy: small microbiome has a big hope. Nano Micro Lett 13:37
66. Akolpoglu MB, Alapan Y, Dogan NO, Baltaci SF, Yasa O, Tural GA et al (2022) Magnetically steerable bacterial microrobots moving in 3D biological matrices for stimuli-responsive cargo delivery. Sci Adv 8:eabo6163 [Internet]. American Association for the Advancement of Science. Available from: https://www.science.org/doi/10.1126/sciadv.abo6163. Cited 6 Oct 2022
67. Shuwen H, Yifei S, Xinyue W, Zhanbo Q, Xiang Y, Xi Y (2024) Advances in bacteria-based drug delivery systems for anti-tumor therapy. Clin Transl Immunol 13:e1518 [Internet]. Wiley. https://doi.org/10.1002/cti2.1518
68. Jackson RJ, Ramsay AJ, Christensen CD, Beaton S, Hall DF, Ramshaw IA (2001) Expression of mouse Interleukin-4 by a recombinant ectromelia virus suppresses cytolytic lymphocyte responses and overcomes genetic resistance to mousepox. J Virol 75:1205–1210 [Internet]. American Society for Microbiology. Available from: https://pubmed.ncbi.nlm.nih.gov/11152493/. Cited 21 Sep 2020
69. Gumuskaya G, Srivastava P, Cooper BG, Lesser H, Semegran B, Garnier S et al (2024) Motile living biobots self-construct from adult human somatic progenitor seed cells. Adv Sci 11:2303575 [Internet]. Wiley. https://doi.org/10.1002/advs.202303575
70. Acevedo-Rocha CG, Budisa N (2016) Xenomicrobiology: a roadmap for genetic code engineering. Microb Biotechnol 9:666–676 [Internet]. Wiley. Available from: https://pubmed.ncbi.nlm.nih.gov/27489097/. Cited 21 Sep 2020
71. Hua Y, Dai X, Xu Y, Xing G, Liu H, Lu T et al (2022) Drug repositioning: Progress and challenges in drug discovery for various diseases. Eur J Med Chem 234:114239
72. Sharma A, Farouk IA, Lal SK, Martinez-Sobrido L, Toral FA (2021) COVID-19: a review on the novel coronavirus disease evolution, transmission, detection, control and prevention. Viruses 13:202 [Internet]. Multidisciplinary Digital Publishing Institute. Available from: https://www.mdpi.com/1999-4915/13/2/202/htm. Cited 20 Apr 2022
73. Kim JR, Hayes KA, Yeung NW, Cowie RH (2013) Definitive, intermediate, paratenic, and accidental hosts of *Angiostrongylus cantonensis* and its molluscan intermediate hosts in Hawai'i. Hawai'i J Med Public Health 10
74. Anderson PD (2012) Bioterrorism: toxins as weapons. J Pharm Pract 25:121–129
75. Christie RJ, Findley DJ, Dunfee M, Hansen RD, Olsen SC, Grainger DW (2006) Photopolymerized hydrogel carriers for live vaccine ballistic delivery. Vaccine 24:1462–1469
76. Schaufelbühl S, Florquin N, Werner D, Delémont O (2024) The emergence of 3D-printed firearms: an analysis of media and law enforcement reports. Forensic Sci Int Synerg 8:100464
77. Casadevall A, Pirofski L-A (2006) The weapon potential of human pathogenic fungi. Med Mycol 44:689–696 [Internet]. Available from: http://www.ncbi.nlm.nih.gov/pubmed/17127625. Cited 22 Mar 2019
78. Block S (2001) The growing threat of biological weapons. Am Sci 89:28–37
79. Elgabry M, Nesbeth D, Johnson SD (2020) A systematic review of the criminogenic potential of synthetic biology and routes to future crime prevention. Front Bioeng Biotechnol 8:571672
80. Park S (2013) DIY biology: bio-inspired or biohazard? Microbiol Today 40:120–123
81. Carnes C. Soviet intelligence support to international terrorism. Am Intell J 7:18–23 [Internet]. National Military Intelligence Foundation. Available from: http://www.jstor.org/stable/44326123
82. Chatzis I, Rompolas P, Karachristos C, Grivopoulou A (2023) Cyberbiosecurity: the threat of the sages. In: Kambouris ME (ed) Biosecurity in the making, 1st edn. CRC Press, Boca Raton, pp 119–132
83. Özdemir V (2018) The dark side of the moon: The Internet of Things, Industry 4.0, and The Quantified Planet. Omi A J Integr Biol 22:637–641 [Internet]. Mary Ann Liebert, Inc., New Rochelle, NY, USA. Available from: https://www.liebertpub.com/doi/10.1089/omi.2018.0143. Cited 10 May 2019

84. Reed JC, Dunaway N (2019) Cyberbiosecurity implications for the laboratory of the future. Front Bioeng Biotechnol 7:182 [Internet]. Frontiers Media S.A. Available from: http://www.ncbi.nlm.nih.gov/pubmed/31497596. Cited 18 Sep 2020
85. Borges do Nascimento IJ, Pizarro AB, Almeida JM, Azzopardi-Muscat N, Gonçalves MA, Björklund M et al (2022) Infodemics and health misinformation: a systematic review of reviews. Bull World Health Organ 100:544–561
86. Bernard R, Bowsher G, Sullivan R (2020) Cyber security and the unexplored threat to global health: a call for global norms. Glob Secur Heal Sci Policy 5:134–141
87. Murch RS, So WK, Buchholz WG, Raman S, Peccoud J (2018) Cyberbiosecurity: an emerging new discipline to help safeguard the bioeconomy. Front Bioeng Biotechnol 6:39
88. Smith TC (2010) Denialism: how irrational thinking hinders scientific progress, harms the planet, and threatens our lives. Emerg Infect Dis 749–750
89. Nawaz R, Thompson P, Ananiadou S (2013) Negated bio-events: analysis and identification. BMC Bioinform 14:14. https://doi.org/10.1186/1471-2105-14-14
90. Godulla A, Seibert D, Klute T (2024) What is denialism? An examination and classification of definitional approaches and relevant actors. J Media 135–147
91. Musk E (2019) An integrated brain-machine interface platform with thousands of channels. J Med Internet Res 21:e16194
92. Ienca M, Scheibner J (2020) What is neurohacking? Defining the conceptual, ethical and legal boundaries. In: Bárd I, Hildt E (eds) Developments in neuroethics and bioethics. Elsevier Academic Press, pp 203–231
93. Brown S (2006) Stealth sharks to patrol the high seas [Internet]. New Sci. Available from: https://www.newscientist.com/article/mg18925416-300-stealth-sharks-to-patrol-the-high-seas/
94. Kambouris ME (2024) The GCBR after COVID-19. Hybrid Warf 22, 1st edn. Springer, Cham, pp 143–186
95. Roth G (2015) Convergent evolution of complex brains and high intelligence. Philos Trans R Soc London Ser B, Biol Sci 370
96. Sato H, Maharbiz MM (2010) Recent developments in the remote radio control of insect flight. Front Neurosci 4:199
97. Nikolić SS (2017) An innovative response to commercial UAV menace-anti-UAV falconry. Vojn Delo 4:146–167 [Internet]. Available from: http://strangevehicles.greyfalcon.us/NaziPigeon.htm. Cited 5 Aug 2022
98. Perry T (2015) Dolphins' and Sea Lions' important role in guarding against terrorism. Gov Technol [Internet]. Available from: https://www.govtech.com/em/safety/Dolphins-Sea-Lions-Train-Navy-Deployment.html. Cited 10 Dec 2020
99. Kambouris ME, Manoussopoulos Y, Velegraki A, Patrinos GP (2023) The biote-bot hybrid. The ultimate biothreat merging nanobots, AI-enabled cybernetics and synthetic biology. Futur Med AI 1:FMAI4 [Internet]. https://doi.org/10.2217/fmai-2023-0008
100. Cieslak TJ, Kortepeter MG, Wojtyk RJ, Jansen HJ, Reyes RA, Smith JO (2018) Beyond the dirty dozen: a proposed methodology for assessing future bioweapon threats. Mil Med 183:e59–e65 [Internet]. Available from: https://pubmed.ncbi.nlm.nih.gov/29401327/. Cited 22 Apr 2022

Chapter 7
AI-Driven Disruptive Biosecurity: Possible Applications

Abstract The Disruptive Biosecurity is meant to tackle the Total Biothreat, but also more elaborate schemes that amplify the Total Biothreat to Meta-Biothreat status. The Disruptive Biosecurity has to be adaptable to respond to the sum of possible, not just probable, threats. These include: conventional bioevents perpetrated by conventional or not bioagents; the agentless events, where standard biosecurity concepts, means and methods and Medical Countermeasures are irrelevant; the cyber-amplified and -engineered bioevents, where biovigilance might have blind spots, despite the implication of real, live bioagents; the massive, surge outbreaks, simultaneous or successive might these be, which may overload response, exhaust personnel, wear equipment and deplete resources. Although response, bioresillience and biodefense were invented, deployed and to some degree still operating without AI, the latter is a toolkit that might unlock performance levels and capabilities essential for response adaptation and preparation for the infinite number and variety of threat scenaria. It may sort, assess, promote and support in planning, operational, logistics and cognitive terms different modes, levels and commodities for biosecurity, including surveillance, diagnostics, protection, prophylaxis and therapeutic amenities. Moreover, its use for Biosecurity might be the sole antidote to balance its use by prospective, AI-legible bio-offenders.

Keywords Non-lethal bioweapons (NLBW) · Narrow/weak AI · Strong/generic AI · Artificial superintelligence · Neurohacking · Biote-bots · Metahumans · Radiotherapies · Electrotherapies · Magnetotherapies

The Disruptive Biosecurity is meant to tackle the Total Biothreat, as has been the current biosecurity/biovigilance and bioresilience concepts, but also other, more elaborate formats, that amplify the Total Biothreat to Meta-Biothreat status, as seen in

This chapter expands and develops issues identified in the paper "Disruptive biosecurity: A new concept for preparedness and response to the emerging total biothreat landscape" by Kambouris ME et al. (2025) published in Comparative Strategy 1–33, https://doi.org/10.1080/01495933.2025.2565586.

M. E. Kambouris, *Preparing for the Meta-Biothreat: Disruptive Biosecurity*, Advanced Sciences and Technologies for Security Applications,
https://doi.org/10.1007/978-3-032-25690-4_7

Chap. 1. The Disruptive Biosecurity has to be adaptable to the sum of threats: From the agentless events, which render biosecurity and Medical Countermeasures irrelevant, through cyber-amplified and -engineered bioevents, where biovigilance might have blind spots, as detailed in Chaps. 5, 6, to the massive, surge outbreaks that may overload response, exhaust personnel, wear equipment and deplete resources. It is supposed to tackle them by adapting, not by some miracle wand that works in all cases—although that would be more than welcome. To a great degree, the needed performance and skills are achievable, or are to be, only thanks to the advent of Artificial Intelligence. Although response, bioresilience and biodefense were invented, deployed and to some degree still working without it, AI is a tool that might unlock capabilities necessary for adaptation and preparation for the infinite biothreat scenaria, and perhaps the only way to counter its own misuse by prospective offenders.

7.1 Reshuffling Response

The minimal perpetration events as discussed previously in Chaps. 5, 6, namely (i) Agentless bioevents, (ii) Cyber-amplified bioevents, and (iii) Cyber-engineered/perpetrated bioevents, may be tackled, at least partly, by creative application of AI by the authorities, or, more explicitly, by the security community. It seems to be the only toolset, in its existing and, more importantly, in its future iterations—the Artificial Superintelligence [1] being the paradigm for the latter—that could detect risk networks rather than linear risk chains, and identify exploitable liabilities within them. This is the quintessence of the idea of the Disruptive Biosecurity as presented herein. Given that any conceptual/possible risk network is to degrade to a sequential, linear form—a pipeline—upon materialization (Fig. 7.1), the nodes of the said networks are actionable and proportionally so to the number of their links [2].

Although a holistic approach towards the prospective dangers inherent in the extreme automation is achievable, indicatively through the concept of Industry 5.0 [3], less integrated solutions can be formulated and are bound to be preferred due to cost and simplicity concerns. They include supervising steps in identified risk sites and amenities, such as may be found in the Internet, e.g. tutorials available in the web for malicious applications ("Internet Warding"). The prospective perpetrators may spot such sites in order to follow them, so the surveillance agency may spot them to track users/visitors showing interest in these courses, which imply Capability in the Making, and then associate it with other interests, purchases and travels.

Capability-centric logic pertains the interception of perpetrated bioevents; it is what someone or some group *may* do that informs the interception, not what they *want* to do. Still, too much noise is bound to surround future Black Biology projects [4, 5]; buying potato flakes may be for cooking or for preferential growing some microbial taxa in potato-containing substrates. This noise makes such moves hard to define and classify (many microorganisms love potato flakes, as so do some parties and other

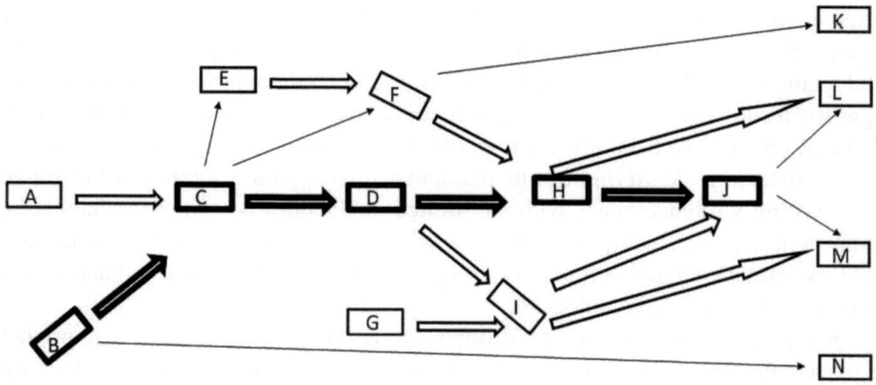

Fig. 7.1 Any conceivable two-dimensional staged risk network upon materialization is to degrade to a sequential, linear form (thick hollow arrows), which, conditionally, may be repeated in space and time and thus consists pipeline. Simple hollow arrows show all possible sequential steps (boxes), from inception (A, B, G) to materialization (K, L, M, N). The solid arrows show steps materialized at any time and in any possible combination among themselves in the past

social events) but easy to detect (there is massive purchase of potato flakes from a single, lone individual not engaging in the food industry). The latter holds if the signal suppression, inherent in biodenialism and bionegationsim [6–8], can be somehow neutralized, filtered or bypassed ("How dangerous can it be to purchase potato flakes, one may be a harmless glutton!"). As a result, surveillance and intelligence resources and algorithms/AI amenities, including medical intelligence, are needed to make sense of all such spotted contacts ("But up to last week he was only purchasing beans, or rice"). To do so better, it might be advantageous to focus on *choke points* of the whole threat process. The latter is meant from inception of an attack/event to the end results and might not be linear/sequential as already mentioned. In such choke points, where multiple plausible pipelines cross and intersect, the methodological and practical solutions for the perpetrator are few and unavoidable and thus uncertainty, and consequently noise, diminish, thus favoring the surveillance authority. Choke points may prove to be (indicatively and not exclusively): (i) the actual possession or acquisition attempts or even access to a set of capabilities and the concomitant issue of spatiotemporal location of the perpetrator and its effectors, which are closely interrelated; and (ii) the pathogenic mechanisms that may be used to cause harm to the biotarget (not necessarily, nor primarily human) and the respective neutralization options, generic or selective may these be.

The approach regarding choke points requires a robust buildup of diagnostic capabilities, so as to use them to survey a target area/population, screen them when a crisis is imminent and also locate possible sources of pathogens (habitats of microorganisms, vectors, locations of ongoing epidemics/endemics). Such capacity would greatly benefit from advanced iterations of medical-oriented AI. Coupled to enhanced diagnostic technologies, even Narrow/Weak AI [9] may improve agent detection (perception of a presence of interest in a controlled entity, may that be individuals or area/volume) and identification (determine the similarity of the entity of interest to any of a number of known, reference entities through comparative quantitative and qualitative scores in relevant biomarkers). The above approach is applicable to agnostic testing [10] so as to detect and identify causative agents initially unknown, that may either be well-known existing, or newly emerging pathogens. It is also relevant to detecting/identifying agents manipulated to bypass detection, for example by rewriting virulence genes with substituted nucleotides so as not to change the aminoacid sequence, while changing the coding DNA sequence. The latter is usually read by sequencing and compared to reference DNA for identification/match, and not referred to encoded aminoacids.

On top of the above, the use of Strong/Generic AI may render possible to predict a given malefactor's selection of agents down to some taxon or functional risk group by considering a set of desired qualities (toxinogenic, highly resistant, highly robust, explosively propagating, airborne, highly persistent etc.). This process is similar to risk projections of today, but it would be carried out with constant updates, perhaps daily. These updates, similarly to Air Tasking Orders of air campaigns, would take the form of a Biosecurity Tasking Order and would include projections for many possible perpetrators and bioagent-vector pairs or rather complexes. Conditions such as accessibility, know-how, technological state-of-the-art are expected to be taken into consideration for formulating the prioritized risk assessment [4, 11, 12].

Such predictive and diagnostic capabilities may assist in resolving an incident, or not, due to the possibly indefinite selection of effectors and vectors. Still, at any given time, they may be able to trim down possibilities and thus *partially* resolve the respective issues, thus allowing targeted diagnostics (existing or *ad hoc* developed by fast-track procedures, including modifications of existing ones) to be used for follow-on, massive screening for either fully identified or for only initially characterized pathogens—with a higher degree of uncertainty in the latter case.

A good example of alternative approaches where prediction is of limited use and agnostic identification may prove more advantageous [10] than specific, targeted diagnostics, is the need to promptly scan for tainted sharp instrument injuries surging in temporal terms but possibly dispersed in spatial terms. This requires AI, even Narrow/Weak AI tools, to associate multi-omics data such as, but not limited to, genomics, proteomics, metabolomics, microbiomics harvested from wound tending stations, be these permanent facilities or *ad hoc* deployed ones. The scan might include not only injuries by mechanical means, but also due to thermal and chemical causality. Multiple biosignatures may be processed at the immediate diagnostics stage, from toxicology and mass spectrometry to metagenomics and culturomics [13], so as to correlate systematic, massive, concerted appearance of infectants, both pathogens and toxins. Common signatures should be inquired in (perhaps minimally) associated incidents, even if the latter are widely dispersed spatiotemporally, in order to shake off surveillance. Furthermore, different infectants might have been used (i.e. cellular agents and/ or molecular agents) to spike the vectors, specifically meant to deny any association conclusions to responder agencies. It is the surge of events, not their nature, identity or *modus operandi* of the agents that should raise the flag at the proper time. Human expertise is indispensable, but under pressure and a tsunami of data retrieved by various sources and updated fast, verging to Big Data values, Strong/Generic AI tools might be an inflexible requirement after all.

The ultimate aspect in preparedness in Biosecurity is preparations and planning for therapeutic—and, conditionally, prophylactic—intervention. It is the most fundamental level; if prevention and protection fail, the syndrome must be contained, at least in terms of symptoms, to stabilize and sustain the patient/health casualty until the immune response builds up and/or other intervention options surface, of causative nature. There is a large but finite number of things to go wrong in a given physiology, as opposed to the prospectively infinite causative agents, if emerging agents (pathogens) and engineered ones (metapathogens) with or without the xenobiology dimension [14–16] are taken into account. Still, the basic disease categories are only four: Infection, neoplasia, injury (mechanical, chemical, thermal) and deregulation/deterioration of spontaneous or congenital causality or due to age and lifestyle.

There are three discrete ways of infectious pathogenicity, which, if combined and multiplied by different levels of tropism (host identity and host's biocompartments) produce the very large but finite number of cases of disease mechanisms. A pathogen, first of all, may trigger a reaction of the immune system, mediated by cytokine storms, resulting in Systemic Inflammatory Response too severe to be managed by the host's homeostasis and causing conditions such as, but not limited to, shock and hemorrhage [17–19]. In this case the pathogen actually does nothing destructive, but remains purely disruptive itself and lets the host system self-destruct by only magnifying, or rather amplifying, the host's own malfunction in regulation of response. Actually, a pathogen is not needed; suffice it to be something simply alien, a xenoagent or xenobiotic moiety [20], possibly benign, that the host immune system would falsely interpret as pathogenic and would react violently to its presence by developing severe inflammatory response. Such cases are the allergens [21].

In most cases, though, a pathogen actively harms the host through the second way: the invasive and degenerative, *septic* pathway, where it degrades tissues in its immediate vicinity and continues to propagate, grow and progress further and/or deeper [22]. This growth may result in dissemination to other biocompartments by spontaneous transportation of invading cells by some element of circulation, such as blood and lymph, although other cases do exist [23]. At the new site, the new focus of the infection again degrades the host tissue by physical contact and continuity, and different tissues may be affected, until a vital organ is injured, and/or shock occurs, if a vital function is corrupted.

And the third possibility is the *toxic* effect. It may be a lone molecule or the product of some cell [23], alien by consensus—although this might need some further exploration as autotoxicosis may be a possibility, and an open window for bioassault in the near future, especially with artificial cells merged within organs, tissues or cells of the host, as discussed in Chap. 4 for the Biocyborgs [24]. The toxic effect by definition affects molecular targets and thus deregulates cell physiology or destroys cells and consequently compromises organs or multi-organ functions. It may impact sites distant from the point of origin with ease, compared to the dissemination of whole cells. But toxic elements do not self-replicate; they are non-replicating molecules, a key prerogative for inclusion in the biochemical, not the pure biological weapons/

agents. By this criterion, prions are an exception [25], if they are considered biotoxins, and the same applies to any pathogenic viroid that may be discovered or designed in the future.

As a result of its non-replicating nature, the quantity and concentration of the toxin are dropping over time and as it diffuses and disseminates. The decay of the active concentration of the toxin may be averted if there is a producer cell/population within the functional limitations of the event that secretes further quantities, as in infections by toxinogenic microbiota [26–28]. Else, the initial dose, if delivered away from a susceptible biocompartment and in low quantity, as in some secondary snake bites, may cause a mild effect or be totally unnoticed [29].

A pathogen may cause any combination of the three: a cell growing by degrading nearby tissues and using the nutrients to grow, multiply and release biotoxins is the paradigm and *B. anthracis* is an example [27, 30–33]. The SARS-CoV-2 bears a partially toxic spike, multiplies in healthy cells and degrades them and causes a cytokine storm. So it is the paradigm for all-three-in-one [17, 34, 35]. Thus, on a conceptual plane, there are less things to do and less unknowns to tackle in terms of pathogenicity than identification of the pathogen. The human physiology is a constant, and bolstering it by supportive care and symptomatic therapy is doable without needing to know the causative agent.

In an event with unknown agents, especially ones produced by Synthetic Biology, there will be no time for designing and producing novel drugs or vaccines: a proper, destruction-oriented bioweapon –even if non-lethal in its nature/NLBW [21]— will wreak havoc long before any such novel amenity is designed, produced and distributed in volume; the COVID-19 vaccines showed that proper testing may be omitted. Thus previous, existing supplies and amenities would have to undergo fast-track evaluation by AI tools, initially Narrow AI until better toolkits of Generic AI become available, which would suggest and prioritize proactively the bulk of expendable medical supplies in this context. The obvious, high-profile duty is of course the repurposing of existing drugs for symptomatic therapies as first-line MCM [36]. Once multi-omics and whole genome analysis reveal molecular targets, Narrow AI tools would be able to screen available antisera and micromolecular compounds that may be effective. Doing so with Biopharmaceuticals may need Wide AI capabilities. The most cost-effective such amenities would be purchased massively for creating emergency stocks. These would be available to routine medical operations so as to have stocks rotated and renewed in order not to become outdated, expired or obsolete. This concept has been known as User-Managed Inventory (UMI) for quite some time [37], but is perfectly valid in the new framework and the abovementioned uses of AI will bolster its implementation.

The above criteria directly suggest the need for economic evaluation; possibly not in monetary terms but rather in terms of producibility and logistics, including supply chain management [38, 39]. Industry 4.0 could be of great assistance by definition [40], but even without its application, Narrow/Weak AI might assist in optimization of the selection procedure. This sector does not concern only drugs, such as anti-inflammatory agents to alleviate cytokine storms and allergies [19]. It includes any supportive means, indicatively and not exclusively respirators, defibrillators and plastic, expendable masks, gloves and any other clinical or laboratory supplies all the way to serum and blood plasma or other blood products and transfusion supplies.

A step before the last resort—and least uncertainty—measures, as are treatment modalities, especially symptomatic therapies and life-support interventions, come the interventions focusing on the causative agent. As there are no reliable indicators in the age of Synthetic Biology and Exo-/Astro-/Xenobiology [41], to determine the stockpiles of antimicrobials is rather conjectural and a privileged information. Knowledge on the subject would inform the perpetrator to select a different vector-effector system, one that would be unaffected by the stockpiled amenities. In this case the notion of vaccines is counterproductive: it takes time to volume-produce a safe and effective vaccine as obvious with complications with the SARS-CoV-2 endeavor [42], while building stockpiles means that the threat agent must be reliably known and stable. For unknown and unstable agents (the latter meaning spontaneous shift of their determinants) or many agents "outbreaking" seemingly chaotically, the massive vaccination is not the proper approach as stocks become obsolete. At the same time, the conditional need for exceptional transportation and stockpiling arrangements, such as cold chain, degrades the sustainability and efficiency of vaccination as an intervention option [43]. It is the one approach with the least expected applicability.

The way to go is obviously *therapeutic* medicaments—or amenities in general—preferably applicable also as prophylaxis [44]. This means that the antimicrobials, understood as different classes of antibiotics but also antimicrobial peptides (AMPs), antiseptics, sterilants and disinfectants [45–47] are valid candidates. An additional—but not vital—aspect, the immunomodulating functionality, as possibly embedded in the use of Biologicals, would increase the differentiation, diversification and thus sustainability and robustness of the response effort [48, 49]. The antimicrobials have the additional advantage of possible cross-efficacy against different microbial taxa and thus may be used also in routine prescription, enabling management and rotation of the stocks and so allow renewal of said stocks by, say, every two or five years. This possibility increases sustainability and affordability, since the stockpiles earmarked for immediate, first-line MCM will not be discarded upon *expiration* date.

As mentioned, cross-reactions allow possible repurposing of existing drugs to future threats. This implies a ready supply of existing, circulating or not, chemotherapeutics, to be used as prompt, first-line MCM upon detection of an outbreak and characterization of the agent [36, 50]. Which of the antimicrobial drugs and other amenities have the best prospect for future, uninformed expedient use (especially prophylaxis) and may worth the expense to be stockpiled is a matter of AI-supported decisions. It must be stressed that the same applies, with the added need for regular updates, to prioritizing developmental drugs as promising for such expediencies and thus inclusion of them in push-funding for expedient development [51, 52] and ability to surge their production upon request, as with the Options Market for Antibiotics (OMA) [53]. Obviously, more generic formulations, with efficacy against groups of microbiota rather than against specific species or strains, such as broad-spectrum antimicrobials (once again taken as antiseptics, disinfectants, AMPs and antibiotics) would be better-positioned [52] despite questions of ecological damage and propagation of resistance to antimicrobials [49, 54–56].

7.2 Innovative Intervention

A somewhat innovative option in causal therapy (although of occasional applicability in symptomatic therapy as well) may be found within the non-expendable amenities, including but not limited to instrumentation for therapies based on different effects and forms of energy. Such are the electrotherapies, magnetotherapies, radiotherapies, sonication/ultrasound treatment and thermotherapies [57–61]. Such amenities, may they be already established or experimental, should be prioritized as they require only re-tuning for any novel pathogen; not novel production lines and extensive testing as with chemopharmaceuticals and biopharmaceuticals. This would be clearly one privileged field for AI applications in two levels: first to prioritize and select methods and protocols, and then to attune these existing or *de novo* amenities, as some cellular/particle qualities may suggest a level of consistency in the optimal settings for a given group of pathogens, even if the actual pathogen is not already properly identified, analyzed and characterized. Thus, in quasi-agnostic cases they would be a valid, affordable and sustainable option.

The other way round, it may be feasible to *predict* the susceptibility of different bioagents, taken individually rather than grouped in taxa, to non-pharmaceutical interventions, such as electroceuticals, mild and intense thermal treatment, pressure application and sonication and perhaps even suggest probable optimal settings [57–61]. If so, it might be feasible to develop predictive algorithms so as to update regularly, with new entries, any list of suggestive intervention settings per microbiote biomarker(s)—implying association to known liabilities and susceptibilities—once the analysis of emerging/novel ones is completed. Moreover, any such capability would render electroceuticals [62–66] and other energy-based therapies prescribable and safely and easily administered in routine cases, further alleviating antibiotics stewardship concerns [67] and increasing the scope of standard antimicrobial treatment options [68, 69].

The abovementioned pathway may be useful in determining (and perhaps compile priority lists by probability regarding) the applicability of diagnostic tests for such groupings of agents. The latter might lead to a tripartite association of (i) biomarkers with (ii) routine, screening diagnostic technologies and (iii) binary or differential scores of susceptibilities to different antimicrobial amenities, predicted by Superintelligent AI once the analysis/characterization of a causative or prospective pathogen is complete.

Additional AI applications against biosecurity challenges may be identified beyond the medicinal context. One such aspect obviously refers to the resolution of possible virtual (as opposed to actual) bioattacks (or *agentless bioattacks*) implemented, as mentioned in Chaps.5, 6, mostly but not entirely by cyber operations [70, 71] and intended mainly to disrupt. Further such applications may include the most worrisome projected threat, which comes from the fully biointegrated Man–Machine interface that opens new threat highways: *metahumans* [72, 73] enhanced with neurotransplants individually wired to neurons may control solely by thought a range of IT-dependent items in the near future, with the FPV (First Person View) drones being a cause for alarm. The concept has been proven since early 2024 by Neuralink [74]. In a more elaborate setting this task may be accomplished either directly or indirectly, through some kind of signal repeaters. The objects of such remote control may be active or hibernating/waiter drones of different types and sizes designed or modified to deliver bioagents, with extreme such case the biote-bots [75], and IDDs/ICDs (as defined in Chap. 6), fixed or moving. To the above, one may add the control through hacking of medical transplants such as neurotransplants—the definition of neurohacking [76]—but also of more mundane items, such as insulin pumps, and pacemakers. All these may be hacked and either controlled, or corrupted or outright deactivated by such *meta*-willpower and perhaps this may extend to medical instrumentation, the example being software-dependent/controlled diagnostic devices or external life-support [77].

Similarly, chipped/wired macroorganisms, mostly animals, might be neurohacked [76], foreseeably to become remotely controlled so as to carry out orders. These range from demonstrating extreme aggression or shifting to swarm/pack behavioral mode to company animals or wildlife, such as sharks and insects [78–80], to guiding flocks of birds residing around airports or flying near flight paths so as to crash *en masse* into civil or military aircraft, especially their air intakes and cockpit canopies, thus bringing them down [81].

In all these cases AI may either assist in scanning the airwaves and intercepting the actual control/hacking code, or selectively activate protective routines embedded locally. This may be either within the operational systems, regarding limited connectivity devices, or, for connected devices, such routines may be available through Cloud Storage [82]. Such routines would be triggered upon detection of markers suggesting hacking so as to neutralize the malevolent code, and, thanks to Wide AI, they may be threat-adaptable, allowing a much more robust cyber-immunity than preconceived anti-malware solutions.

Regarding the control and neutralization of dispersion apparatuses, such as, but not limited to ICD/IDD, the Wide AI would mimic military Electronic Warfare (EW) operations [83]: prompt scanning of the airwaves, hacking the command and control codes to deny proper transmission/reception of commands (or transmission of own commands instead, such as self-destruct, or relinquishing control) and, if unable to do the latter, simply interrupting the link between controller and device. This is achievable by either volume jamming or by hacking into the digital conversation and denying service; or by targeted noise-jamming of the reception, which may lead to any from a number of suppressive options such as aborting and auto-returning to idle/standing down, or self-destructing. In all, AI will be the indispensable piece of the puzzle for countering some cyberbiothreats and an invaluable one, although not the only one, for tackling or containing other biothreats.

References

1. Brödner P (2018) "Super-intelligent" machine: technological exuberance or the road to subjection. AI Soc 33:335–346 [Internet]. https://doi.org/10.1007/s00146-017-0731-6
2. Manoussopoulos Y, Anastassopoulou C, Ioannidis JPA, Tsakris A (2023) Paired associated SARS-CoV-2 spike variable positions: a network analysis approach to emerging variants. mSystems 8:e0044023
3. Özdemir V, Hekim N (2018) Birth of Industry 5.0: making sense of big data with artificial intelligence, "The Internet of Things" and Next-Generation Technology Policy. Omi A J Integr Biol 22:65–76 [Internet]. Available from: http://www.liebertpub.com/doi/10.1089/omi.2017.0194. Cited 10 May 2019
4. Casadevall A, Pirofski L-A (2006) The weapon potential of human pathogenic fungi. Med Mycol 44:689–696 [Internet]. Available from: http://www.ncbi.nlm.nih.gov/pubmed/17127625. Cited 22 Mar 2019
5. Block S (2001) The growing threat of biological weapons. Am Sci 89:28–37
6. Smith TC (2010) Denialism: how irrational thinking hinders scientific progress, harms the planet, and threatens our lives. Emerg Infect Dis 749–750
7. Nawaz R, Thompson P, Ananiadou S (2013) Negated bio-events: analysis and identification. BMC Bioinform 14:14 [Internet]. https://doi.org/10.1186/1471-2105-14-14
8. Godulla A, Seibert D, Klute T (2024) What is denialism? An examination and classification of definitional approaches and relevant actors. J Media 135–147
9. Korteling JEH, van de Boer-Visschedijk GC, Blankendaal RAM, Boonekamp RC, Eikelboom AR (2021). Human- versus artificial intelligence. Front Artif Intell 4:622364
10. Leiser OP, Hobbs EC, Sims AC, Korch GW, Taylor KL (2021) Beyond the list: bioagent-agnostic signatures could enable a more flexible and resilient biodefense posture than an approach based on priority agent lists alone. Pathogens 10:1497 [Internet]. Multidisciplinary Digital Publishing Institute. Available from: https://www.mdpi.com/2076-0817/10/11/1497/htm. Cited 6 Oct 2022
11. Cieslak TJ, Kortepeter MG, Wojtyk RJ, Jansen HJ, Reyes RA, Smith JO (2018) Beyond the dirty dozen: a proposed methodology for assessing future bioweapon threats. Mil Med 183:e59–e65. Available from: https://pubmed.ncbi.nlm.nih.gov/29401327/. Cited 22 Apr 2022
12. Casadevall A, Pirofski L (2004) The weapon potential of a microbe. Trends Microbiol 12:259–263 [Internet]. Available from: http://www.ncbi.nlm.nih.gov/pubmed/15165603. Cited 22 Mar 2019
13. Velegraki A, Zerva L (2020) Identifying microbiota: genomic, mass-spectrometric, and sero-diagnostic approaches. In: Kambouris M, Velegraki A (eds) Microbiomics: dimensions, applications, and translational implications of human and environmental microbiome research, 1st edn. Elsevier Academic Press, London, pp 77–94
14. Acevedo-Rocha CG, Budisa N (2016) Xenomicrobiology: a roadmap for genetic code engineering. Microb Biotechnol 9:666–676 [Internet]. Wiley. Available from: https://pubmed.ncbi.nlm.nih.gov/27489097/. Cited 21 Sep 2020
15. Schmidt M, Pei L, Budisa N (2018) Xenobiology: state-of-the-art, ethics, and philosophy of new-to-nature organisms. Adv Biochem Eng Biotechnol 162:301–316. Springer Science and Business Media Deutschland GmbH
16. Diwo C, Budisa N (2018) Alternative biochemistries for alien life: basic concepts and requirements for the design of a robust biocontainment system in genetic isolation. Genes (Basel) 10 [Internet]. MDPI AG. Available from: http://www.ncbi.nlm.nih.gov/pubmed/30597824. Cited 21 Sep 2020
17. Hu B, Huang S, Yin L (2021) The cytokine storm and COVID-19. J Med Virol 93:250–256 [Internet]. Available from: https://pubmed.ncbi.nlm.nih.gov/32592501/. Cited 20 Aug 2022
18. Weiss C, Carriere M, Fusco L, Fusco L, Capua I, Regla-Nava JA et al (2020) Toward nanotechnology-enabled approaches against the COVID-19 pandemic. ACS Nano 14:6383–6406 [Internet]. American Chemical Society. Available from: /pmc/articles/PMC7299399/. Cited 29 Nov 2022

19. Icenogle T (2020) COVID-19: infection or autoimmunity. Front Immunol 11 [Internet]. Available from: https://pubmed.ncbi.nlm.nih.gov/33042116/. Cited 20 Apr 2022
20. Esteve-nunez A, Caballero A, Ramos JL (2001) Biological begradation of 2, 4, 6-Trinitrotoluene. Microbiol Mol Biol Rev 65:335–352
21. Galatas I (2017) The misuse and malicious uses of the new biotechnologies. Ann des Mines - Réalités Ind 2017:103–108 [Internet]. Institut Mines-Télécom. Available from: https://www.cairn.info/load_pdf.php?ID_ARTICLE=RINDU1_171_0103
22. Van Baarlen P, Van Belkum A, Summerbell RC, Crous PW, Thomma BPHJ (2007) Molecular mechanisms of pathogenicity: how do pathogenic microorganisms develop cross-kingdom host jumps? FEMS Microbiol Rev 31:239–277 [Internet]. Available from: https://pubmed.ncbi.nlm.nih.gov/17326816/. Cited 30 Apr 2022
23. Anderson PD, Bokor G (2012) Bioterrorism: pathogens as weapons. J Pharm Pract 25:521–529 [Internet]. SAGE Publications Inc. Available from: https://pubmed.ncbi.nlm.nih.gov/23011963/. Cited 22 Feb 2021
24. Elani Y (2021) Interfacing living and synthetic cells as an emerging frontier in synthetic biology. Angew Chemie Int Ed 60:5602–5611 [Internet]. Wiley. Available from: https://onlinelibrary.wiley.com/doi/full/10.1002/anie.202006941. Cited 6 Oct 2022
25. Verma A (2016) Prions, prion-like prionoids, and neurodegenerative disorders. Ann Indian Acad Neurol 19:169 [Internet]. Medknow Publications. Available from: http://www.annalsofian.org/text.asp?2016/19/2/169/179979. Cited 24 Feb 2021
26. Johnson R, Mylona E, Gad Frankel (2018) Typhoidal salmonella: distinctive virulence factors and pathogenesis. Available from: https://doi.org/10.1111/cmi.12939. Cited 30 Nov 2022
27. Taylor-Joyce G, Manoharan S, Brooker TA, Hernández-Rodríguez CS, Baillie L, Oyston PCF et al (2023) The influence of extrachromosomal elements in the anthrax "cross-over" strain *Bacillus cereus* G9241. Front Microbiol 14:1113642
28. Clarke SC (2005) Bacteria as potential tools in bioterrorism, with an emphasis on bacterial toxins. Br J Biomed Sci 62:40–46
29. Ascoët S, De Waard M (2020) Diagnostic and therapeutic value of aptamers in envenomation cases. Int J Mol Sci 21:3565
30. Pilo P, Frey J (2018) Pathogenicity, population genetics and dissemination of *Bacillus anthracis*. Infect Genet Evol 64:115–125 [Internet]. Available from: https://pubmed.ncbi.nlm.nih.gov/29935338/. Cited 30 Nov 2022
31. Pena-Gonzalez A, Rodriguez-R LM, Marston CK, Gee JE, Gulvik CA, Kolton CB et al (2018) Genomic characterization and copy number variation of *Bacillus anthracis* plasmids pXO1 and pXO2 in a historical collection of 412 strains. mSystems 3
32. Broussard LA (2001) Biological agents: weapons of warfare and bioterrorism. Mol Diagnosis 6:323–333 [Internet]. Churchill Livingstone Inc. Available from: https://pubmed.ncbi.nlm.nih.gov/11774197/. Cited 30 Sep 2020
33. Hart CA, Beeching NJ (2002) A spotlight on anthrax. Clin Dermatol 20:365–375 [Internet]. Elsevier Inc. Available from: https://pubmed.ncbi.nlm.nih.gov/12208624/. Cited 25 Sep 2020
34. Moghaddar M, Radman R, Macreadie I (2021) Severity, pathogenicity and transmissibility of delta and lambda variants of SARS-CoV-2, toxicity of spike protein and possibilities for future prevention of COVID-19. Microorganisms
35. Biering SB, Gomes de Sousa FT, Tjang L V, Pahmeier F, Zhu C, Ruan R, et al. SARS-CoV-2 Spike triggers barrier dysfunction and vascular leak via integrins and TGF-β signaling. Nat Commun [Internet]. 2022;13:7630. Available from: https://doi.org/10.1038/s41467-022-34910-5
36. Hua Y, Dai X, Xu Y, Xing G, Liu H, Lu T et al (2022) Drug repositioning: progress and challenges in drug discovery for various diseases. Eur J Med Chem 234:114239
37. Coleman C, Hrdina C, Casagrande R, Cliffer K, Mansoura M, Nystrom S et al (2012) User-managed inventory: an approach to forward-deployment of urgently needed medical countermeasures for mass-casualty and terrorism incidents. Disaster Med Public Health Prep 6:408–414

38. Pinna R, Carrus P, Marras F (2015) The drug logistics process: an innovative experience. TQM J 27:214–230
39. Tucker EL, Daskin MS (2022) Pharmaceutical supply chain reliability and effects on drug shortages. Comput Ind Eng 169 [Internet]. Pergamon Press, Inc. https://doi.org/10.1016/j.cie.2022.108258
40. Özdemir V (2018) The dark side of the moon: The Internet of Things, Industry 4.0, and The Quantified Planet. Omi A J Integr Biol 22:637–641 [Internet]. Mary Ann Liebert, Inc., New Rochelle, NY, USA. Available from: https://www.liebertpub.com/doi/10.1089/omi.2018.0143. Cited 10 May 2019
41. Acevedo-Rocha CG, Fang G, Schmidt M, Ussery DW, Danchin A (2013) From essential to persistent genes: a functional approach to constructing synthetic life. Trends Genet 29:273–279
42. Orient J (2023) Beyond negative evidence: lessons from the disputes on DNA contamination of COVID-19 vaccines. J Am Physicians Surg 28:106–112. Available from: https://jpands.org/jpands2804.htm
43. Rouphael NG, Paine M, Mosley R, Henry S, McAllister DV, Kalluri H et al (2017) The safety, immunogenicity, and acceptability of inactivated influenza vaccine delivered by microneedle patch (TIV-MNP 2015): a randomised, partly blinded, placebo-controlled, phase 1 trial. Lancet 390:649–658 [Internet]. Lancet Publishing Group. Available from: https://pubmed.ncbi.nlm.nih.gov/28666680/. Cited 10 Dec 2020
44. Green MS, LeDuc J, Cohen D, Franz DR (2019) Confronting the threat of bioterrorism: realities, challenges, and defensive strategies. Lancet Infect Dis e2–e13. Lancet Publishing Group
45. Mehraj I, Hamid A, Gani U, Iralu N, Manzoor T, Saleem Bhat S (2024) Combating antimicrobial resistance by employing antimicrobial peptides: immunomodulators and therapeutic agents against infectious diseases. ACS Appl Bio Mater
46. Vijayakumar R, Sandle T (2019) A review on biocide reduced susceptibility due to plasmid-borne antiseptic-resistant genes—special notes on pharmaceutical environmental isolates. J Appl Microbiol 126:1011–1022
47. Rutala WA, Weber DJ (2015) Disinfection, sterilization, and control of hospital waste. Mand. Douglas, Bennett's Princ Pract Infect Dis 3294–3309.e4
48. Makvandi P, Song H, Yiu CKY, Sartorius R, Zare EN, Rabiee N et al (2023) Bioengineered materials with selective antimicrobial toxicity in biomedicine. Mil Med Res 10 [Internet]. Available from: https://pubmed.ncbi.nlm.nih.gov/36829246/. Cited 29 Jul 2023
49. Flandroy L, Poutahidis T, Berg G, Clarke G, Dao MC, Decaestecker E et al (2018) The impact of human activities and lifestyles on the interlinked microbiota and health of humans and of ecosystems. Sci Total Environ 627:1018–1038 [Internet]. Elsevier B.V. Available from: https://pubmed.ncbi.nlm.nih.gov/29426121/. Cited 21 Sep 2020
50. Barrow P, Dujardin JC, Fasel N, Greenwood AD, Osterrieder K, Lomonossoff G et al (2020) Viruses of protozoan parasites and viral therapy: is the time now right? Virol J 17 [Internet]. Available from: https://pubmed.ncbi.nlm.nih.gov/32993724/. Cited 20 Aug 2022
51. Wasan H, Singh D, Reeta KH, Gupta YK (2023) Landscape of push funding in antibiotic research: current status and way forward. Biology (Basel) 12
52. Billington JK (2015) The ABCs of the US broad spectrum antimicrobials program: antibiotics, biosecurity, and congress. Health Secur 13:349–354
53. Brogan DM, Mossialos E (2013) Incentives for new antibiotics: the Options Market for Antibiotics (OMA) model. Global Health 9:58 [Internet]. https://doi.org/10.1186/1744-8603-9-58
54. WHO Press (2010) Antimicrobial resistance: revisiting the "tragedy of the commons." Bull World Health Organ 88:805–806
55. Dyar OJ, Obua C, Chandy S, Xiao Y, Stålsby Lundborg C, Pulcini C (2016) Using antibiotics responsibly: are we there yet? [Internet]. Future Microbiol 1057–1071. Future Medicine Ltd. Available from: https://pubmed.ncbi.nlm.nih.gov/27501941/. Cited 30 Sep 2020
56. English BK, Gaur AH (2010) The use and abuse of antibiotics and the development of antibiotic resistance. Adv Exp Med Biol 659:73–82 [Internet]. Available from: https://pubmed.ncbi.nlm.nih.gov/20204756/. Cited 30 Sep 2020

57. Hunckler J, de Mel A (2017) A current affair: electrotherapy in wound healing. J Multidiscip Healthc 179–194
58. Stanley SA, Friedman JM (2019) Electromagnetic regulation of cell activity. Available from: http://perspectivesinmedicine.cshlp.org/
59. Joshi S, Mobeen A, Jan K, Bashir K, Azad ZRAA (2019) Emerging technologies in dairy processing: present status and future potential. Health Saf Asp Food Process Technol 105–120 [Internet]. Springer, Cham. Available from: https://link.springer.com/chapter/10.1007/978-3-030-24903-8_6. Cited 30 Aug 2021
60. Aipire A, Mahabati M, Cai S, Wei X, Yuan P, Aimaier A et al (2020) The immunostimulatory activity of polysaccharides from *Glycyrrhiza uralensis*. PeerJ 8 [Internet]. Available from: https://pubmed.ncbi.nlm.nih.gov/32030319/. Cited 20 Apr 2022
61. Kaszuba-Zwoińska J, Gremba J, Gałdzińska-Calik B, Wójcik-Piotrowicz K, Thor PJ (2015) Electromagnetic field induced biological effects in humans. Przegl Lek 72:636–641
62. Kambouris ME, Markogiannakis A, Arabatzis M, Manoussopoulos Y, Kantzanou M, Velegraki A (2017) Wireless electrostimulation: a new approach in combating infection? Future Microbiol 12:255–265
63. Sen CK, Mathew-Steiner SS, Das A, Sundaresan VB, Roy S (2020) Electroceutical management of bacterial biofilms and surgical infection. Antioxid Redox Signal 33:713–724
64. Kambouris M, Zagoriti Z, Lagoumintzis G, Poulas K (2014) From therapeutic electrotherapy to electroceuticals: formats, applications and prospects of electrostimulation. Annu Res Rev Biol 4:3054–3070
65. Stathoulias A, Milioni A, Kritikou S, Karmakolia A, Goudoudaki S, Siamoglou S et al (2020) Toward high-throughput fungal electroculturomics and new omics methodologies in 21st-century microbiology and ecology. Omi A J Integr Biol 24:493–504
66. Spanou G, Daskou M, Kambouris ME, Mitsagga C, Mossialos D, Velegraki A et al (2023) Introducing multifactorial electroculturomics: alternating current electric pulses, combined with mild thermal treatment, exhibit antimicrobial or stimulatory effects on bacterial pathogens and enteroviruses, implying prospects for targeted microbiomics a. bioRxiv [Internet]. 2023.11.24.568545. Available from: http://biorxiv.org/content/early/2023/11/24/2023.11.24.568545.abstract
67. Doron S, Davidson LE (2011) Antimicrobial stewardship. Mayo Clin Proc 1113–1123 [Internet]. Elsevier Ltd. Available from: https://pubmed.ncbi.nlm.nih.gov/22033257/. Cited 30 Sep 2020
68. Ogunrinola GA, Oyewale JO, Oshamika OO, Olasehinde GI (2020) The human microbiome and its impacts on health. https://doi.org/10.1155/2020/8045646
69. Cui L, Morris A, Ghedin E (2013) The human mycobiome in health and disease. Genome Med 5:63 [Internet]. Available from: http://genomemedicine.biomedcentral.com/articles/10.1186/gm467
70. Bernard R, Bowsher G, Sullivan R (2020) Cyber security and the unexplored threat to global health: a call for global norms. Glob Secur Heal Sci Policy 5:134–141
71. Murch RS, So WK, Buchholz WG, Raman S, Peccoud J (2018) Cyberbiosecurity: an emerging new discipline to help safeguard the bioeconomy. Front Bioeng Biotechnol 6:39
72. Chiang T (2000) Catching crumbs from the table. Nature 405:517 [Internet]. https://doi.org/10.1038/35014679
73. Pio-Lopez L (2021) The rise of the biocyborg: synthetic biology, artificial chimerism and human enhancement. New Genet Soc 40:599–619 [Internet]. Available from: https://www.tandfonline.com/action/journalInformation?journalCode=cngs20
74. Drew L (2024) Neuralink brain chip: advance sparks safety and secrecy concerns. Nature 19
75. Kambouris ME, Manoussopoulos Y, Velegraki A, Patrinos GP (2023) The biote-bot hybrid. The ultimate biothreat merging nanobots, AI-enabled cybernetics and synthetic biology. Futur Med AI 1:FMAI4 [Internet]. https://doi.org/10.2217/fmai-2023-0008
76. Elgabry M, Nesbeth D, Johnson SD (2020) A systematic review of the criminogenic potential of synthetic biology and routes to future crime prevention. Front Bioeng Biotechnol 8:571672

77. Reed JC, Dunaway N (2019) Cyberbiosecurity implications for the laboratory of the future. Front Bioeng Biotechnol 7:182 [Internet]. Frontiers Media S.A. Available from: http://www.ncbi.nlm.nih.gov/pubmed/31497596. Cited 18 Sep 2020
78. Sato H, Maharbiz MM (2010) Recent developments in the remote radio control of insect flight. Front Neurosci 4:199
79. Nazir F, Malik M, Hussain M, Asif N (2017) Insects as biological weapons. J Bioterror Biodef 09.
80. Brown S (2006) Stealth sharks to patrol the high seas [Internet]. New Sci. Available from: https://www.newscientist.com/article/mg18925416-300-stealth-sharks-to-patrol-the-high-seas/
81. Burger J (1983) Bird control at airports. Environ Conserv 10:115–124 [Internet]. Cambridge University Press. Available from: http://www.jstor.org/stable/44517477
82. Aslan O, Ozkan-Okay M, Gupta D (2021) A review of cloud-based malware detection system: opportunities, advances and challenges. Eur J Eng Technol Res. 6:1–8
83. Li Y, Liu Q (2021) A comprehensive review study of cyber-attacks and cyber security; emerging trends and recent developments. Energy Rep 7:8176–8186

Chapter 8
Seeking a Ubiquitous Disruptive Pattern Through a Unified Disease Theory

Abstract A Unified Disease Theory, even if incomplete, would revolutionize medical thinking and practice. To re-evaluate Life, especially its States, assigning additional importance to non-cellular and immaterial issues, such as hierarchy, communication, interdependence, exchanges of information and matter and control may be needed. The independently procreating life units exist either individually or in collectives of different levels and principles of organization but usual adaptations include defining roles, assigning tasks and transforming to suit functionality prerogatives. Such collectives show variable tolerance to dissimilar or alien units and appears in a number of forms due to evolution. Tissues, organs and multicellular organisms are further evolutions of the same pattern in more evolved and dedicated, eukaryotic cells, with more elaborate means for communication, coordination and differentiation. And the top state is the different communities, of communicating and coordinately functioning, but not necessarily synchronously living, individuals, separated in spatial terms but relevant and associated in temporal terms. Whether this relevance is also functional and allows consideration of such collectives as (meta)organisms rather than perceiving them as societies/communities is vital for understanding but also for responding to challenges posed to different aspects of health by such entities, ranging from infections to tumors.

Keywords Metaorganisms · Eukaryotes · Prokaryotes · Multicellularity · Multimicrobial biofilms · Colonial organisms · Distributed organisms · Gonochoric lifeforms · Persister cells · Hologenome

The Disruptive Biosecurity scheme is preferentially targeting nodes and thrives in common approaches, optimized as needed for seemingly different conditions. If the multitude of pathogens and diseases could be reduced to a common denominator, however wide or complex, the potential of retasking intervention, meant as a disruptive operation and not in its medicinal and pubic health/environmentalism context, would be maximized. This would be achieved and significantly enabled if the four basic disease categories mentioned in Chap. 7, or at least any number of them, could be unified under a common theory regarding their mechanistic causality.

M. E. Kambouris, *Preparing for the Meta-Biothreat: Disruptive Biosecurity*, Advanced Sciences and Technologies for Security Applications,
https://doi.org/10.1007/978-3-032-25690-4_8

Irrespective of the different levels of such convergence, in the sector of Medical Microbiology, recasting the understanding of organisms regarding their vegetative (*sensu lato*) conformation and structure might offer an innovative, more expanded aspect with inherently increased interactions, nodes in their function and thus application points for disruptive countering the caused diseases, in both Environmental Health and in Biosecurity contexts.

The school view on the classification of the States of Life has been around for centuries and suggests a convenient duality: Unicellular and Multicellular organisms. The former may be prokaryotes or eukaryotes, the latter are eukaryotes. In fact it may not be necessarily so, and the recognition of the viruses as lifeforms introduces the Viral Domain in the Acellular Empire of Life, separate from the hitherto recognized, cellular Empire [1–3]. If evidence hitherto explained away as "traits" or communities is re-interpreted as additional levels of lifeforms, a novel scheme emerges with four, not two levels, and greater complexity. This 4-level scheme of classifying Life might include cancer and provide the basis for establishing a Unified Disease Theory, which, if correct, even if not containing all kinds of ailments and diseases, it may prompt the development of not just additional, but perhaps entirely novel prophylactic and therapeutic amenities for diverse diseases.

The most important issue is the change of focus from material to immaterial entities (immaterial being a relative term, of course): focusing away from the cells and organisms and rather onto their relations and interactions, connectivity and interdependence, the abilities that make the Whole much more than the sum of its parts. If these alleged interactions can be targeted and disrupted, and thus seriously compromised or even—ideally—severed, probably the infectious potential would be greatly diminished. The prospect holds tremendous biosecurity potential, as it offers more opportunities for disruptive interventions in the biological constituent of the biothreat, along with the obvious one of public and environmental health.

8.1 A Matter of Resilience

In a biosecurity context, cancer usually receives minimal attention. It is an internal affair of an organism and it takes time to produce health casualties. It is unpredictable and unstable, making it a very unreliable instrument. Still, it is a perfect weapon for long-term results, from democide of a given population exposed to some or other trigger of carcinogenesis (like chemical agents [4], radiological entities [5], electromagnetic radiation [6] but also biological agents, as with highly oncogenic viruses [7]) to more targeted, individual applications of the same principles (as in the fabled assassinations by cancer [8]). The latter provides deniability and a sense of terror and impotence, making the fear of it an ultimate weapon of terror and a leverage provider in the shadow world. If the above are combined with the infectious aspect of perpetrated bioassault, it becomes clear that cancer and microorganisms make a substantial fraction of the total biothreat, especially in more traditional terms, whence macroorganisms and agentless threats are excluded. Even toxicosis/poisoning/bioregulation may be included, as cancer is a dysregulation and many infections have a toxic/dysregulative effect as well as the septic one, the obvious examples being the Anthrax, with its plasmid-encoded toxin and SARS-CoV-2 with the part of the spike bearing homology to bungarotoxin [9–12]. A cognitive scheme that would unify cancer and infection would allow a substantial improvement in Bioresilience: preventive and therapeutic agents might be designed to address both cancer and infection, at least a part of infections with causative agents similar to cancer cells, which is taken to mean eukaryotes. But this may be too short-sighted [13]; other common areas and thus liabilities may be identified in time [14]. As a result, stockpiles with amenities that are applicable in both or multiple occasions may be stockpiled more efficiently and also rotated to alleviate expiration date burdens for such supplies. But beyond that, non-pharmaceutical amenities can be targeted to similar cellular liabilities, or procedural ones, regarding the onset and development of an ailment: thus, the highest impact is expected by cognitive and immaterial, not pharmacological commonality in addressing both cancerous and infectious threats. Taking the issue a step further, opportunities may arize to further disrupt infectants, cancer included (a self-infection aspect of cancer [15] is a valid approach to treating it) if hitherto unknown levels of organization, communication, coordination and interaction are discovered. Given that such structures, functional if not physical, would have been evolved to increase survivability, fitness in general and infectivity/pathogenicity, impairing them in any of the above cases is expected to degenerate the detrimental effect of said pathogens in terms of magnitude, impact and duration.

8.2 The Basics of Life

The fundamental classification of the organization of life has been extremely resilient, despite the enormous progress in every field of Biology since its inception. Independent life is the ability of a life unit to survive, develop (possibly grow) and procreate without involvement of other similar or dissimilar beings. The most basic such manifestation lies with the unicellular, nucleate or not, lifeforms. The pair unicellular-multicellular organism still seems unchallenged, although the increasing consensus on the status of viruses as living entities clearly creates the *acellular* category [16, 17]. This is a first indication that integrative biology is likely to reconsider the original discrimination. For starters, if the viruses are included, the cellular prerogative of life, that life equals cells, no longer holds water. If viruses are lifeforms [1, 18], one would wonder why not viroids, which are single-molecules (contrary to the large number of molecules that constitute a virus) but actually act similarly to the viruses [19]; they simply discarded of the capsid and other protein accessories, making their lives simpler. And then, others may follow, possibly beyond the carbon chemistry—part of which is defined as Biochemistry—[16]; perhaps the crystals [20, 21] or celestial bodies [16]. The Astrobiology has an infinite field of relevance [22–24], although planetary science and chemistry may be, at the time, better poised to tackle the task than outcrops of the earth-attuned biology [25], with the possible exception of Xenobiology [26].

Within the realm of biology, it has been suggested that the cellular-centric concept might be obsolete or even erroneous. A genome-centric concept seems possible if not probable [27]. The idea that the cell is nothing more for its genome than the capsid for the virus—i.e. its vector, vehicle or vestige—is not without merit. It may even be that the cell is the counterpart of the *technida* for the humans and other species performing feats of engineering, including but not limited to ants, termites, beavers, wasps and bees [28], but with a much higher integration; a million-years evolution equaling what modern technoscience defines as "man–machine interface". Should the hypothesis of Bacterial Onset of Cancer Cells-BOCC [29] have some merit, which is corroborated by the perinuclear location of endotumoral intracellular bacteria [30–32], the genome-centricism would become more plausible, as the downloading of genomic content would result in a cell being reprogrammed to change species and domain.

8.3 The Structural Basis of a Unified Disease Theory

There are two possible slots to revise the original couple of organism complexity: the one is to introduce between the unicellular and the multicellular States of Life another one, the *colonial*. The other is to introduce a state beyond the multicellular. It may be called a *collective* [33], a *swarm* [34] or a *hive* [35]; or a *meta-organism* [36], the last being very catchy. It recasts as *organisms* entities that once were considered *communities* [37], a task and process facilitated by the integrative biology concepts such as the holobiont [38–40], the microbiome [41] and all the-ome collectivities, starting from the advent of the genome as a cognitive entity [42].

8.3.1 The Cancer as a (Self-)Predatory Lifeform

The development of cancer may be described as an autoparasitic infection of renegade cells turning autoinfectious in a septic sense. In its simplest form, it regards the malignant independence of one cell from the central control and its abuse of the available resources of the organism, so as to grow and proliferate uncontrollably, with its morphology and ontogenetic form-function duality (as enacted in the context of multicellularity) either subverted or totally nullified. This re-unicellullarization, expressed as both loss of differentiation and precedence of individuality over collectivity is known as the "atavism hypothesis of cancer" [43]. Tumors have considerable genetic heterogeneity [44], and this may be due to the genetic instability of the malignant cells, making the spawn different from the parental ones; or due to the processes inherent in the BOCC hypothesis [29], which will be visited upon later; or due to the fact of multiple events, forming multiple lesions, which diffuse and perhaps intertwine unpredictably within a tumor. The germlines of tumorous cells are clonal [45, 46], and create a lesion, which may grow into a tumor by itself, or, while growing, merge with other lesions and collectively form a tumor. In fact, the constitution of a tumor by many lesions of different genomic context suggests that each lesion is a separate event and as it grows, it comes to contact with other lesions and create the counterpart of a multispecies biofilm [47].

The tumorigenesis in the cellular, or rather nuclear level may be spontaneous, but this is rather rare and within the provisions of the immune system, innate or adaptive, to counter [48]. Generally it follows a cell transformation event that may be triggered by microbial agents, especially viruses [32, 49], but bacteria seem to gain interest, either as facilitators through their effect to the host genomes, cells, tissues and systems [32, 50–53], or directly, by actually creating cancer cells by hacking the genomes of aged (senescent) or otherwise compromised host cells and downloading (parts of it) to their own (or vice versa), as proposed by the BOCC hypothesis [29]. Another causal pathway would be by bio-/chemo- or radiotoxins, agents and amenities, radiation included [4, 5, 46, 52]. It may be the result of the compromise, occasionally to the point of failure, of the immune system which normally tackles the spontaneous cell transformation to malignant forms [54]. It consists of a deregulation of one or more cells regarding the expression of their genomic information and the intensification/re-initiation of their propagation—among other things [46]. The abovementioned pair of conditions may be viewed as a failure in the biochemical and molecular networking of differentiated and purpose-evolved cells in a multicellular context [43]. Still, these two events lead to a tumor directly and indirectly, i.e. by the causative accumulation of other carcinogenesis hallmarks, as the resistance to commanded or spontaneous apoptosis [43, 46, 55, 56]. Thus, a community of relevant and actually kindred cells diverting from the regulation embedded in their position and role within the parent organism [46]. Their further sojourn might be parallelized to an infection, as they absorb resources to the injury of the intact tissues into which they invade [46, 55]—and, as a result, the depletion and exhaustion of the host organism occurs [57].

The ontogenetic deregulation—or dysregulation—seems to impact the cellular physiology proper by detuning internal, cellular functions [46, 55]. Although unicellular eukaryotic organisms have fully functional mitochondria [55], in tumors the Warburg effect is observed—not unlike microbial infections: glycolysis is the main energy source, a fast but quite inefficient process compared to oxidative phosphorylation which explains the insatiable need of the tumor for carbohydrates-especially glucose. The local competition for glucose weakens and neutralizes, through glucose deprivation, the immune cells, which are further contained by suppressive or toxic glycolytic byproducts [55]. As a result, primary energy sources, especially glucose, are withheld by the tumor, which is abundantly provided by blood due to its induction of angiogenesis, resulting in the depletion and subsequent starvation of the tissues of the macroorganisms [57], causing failure of different functions and organs.

It could be proposed that the mitochondria turn superfluous or rather cannot be managed when central command and coordination are switched off [14, 58]. This may refer to an energy demand high enough to surpass the possible mitochondrial output; or to the inability of the cell to sustain their integrity by producing mitochondrial proteins encoded in the nuclear genome, due to the derailment of genomic machinery and production; or to local, mitochondrial pipelines being disturbed, such as (indicatively but not exhaustively) the movement of metabolites.

Another important feature is that tumor cells are not subject to density-dependent inhibition. Normal cells proliferate in culture until they reach a finite cell density, at which point they become quiescent. Tumor cells, however, continue to proliferate independent of cell density [46]. This begs more and deeper examination. This kind of massive, unrestricted growth of conspecific cells is reminiscent of *clumping*, i.e. conspecific members amassing close together, in physical but not functional contact [59]. It suggests lower order of blueprint or total lack of it, even by the lax standards of colonial organisms: the colonial development of bacteria reaches a density-awareness stop, whence the cells become quiescent even if the local density diminishes; by subtracting biomass; on the contrary, extracted biomass moved to fresh substrate, initiates afresh development, suggesting an quorum –sensing amenity, possibly a restrictive metabolite secreted *in-situ* which functions as an inhibitive signal [60].

Although cancer pathology is much more complicated than these few words, there are some more parallels with a microbial infection: tumors do follow rules and blueprints in their development; simply not the ones of the macroorganism, and their rules may be less stringent than the latter ones. For example, a tumor reproduces; it does not only grow, but tumor cells may migrate, actually disseminate, and initiate one or more additional tumors; in other words, they produce additional foci of cancerous infection, at another biocompartment altogether or at another position in the same biocompartment. It is the *metastasis* event [44, 46].

Furthermore, within the tumor, persister cells may be observed, a direct counterpart to the persistence and hypobiosis phenomena amongst the microbiota [61], which ensure their survival under duress [13, 62, 63]. The differences between bacterial/microbial persisters and tumors' ones, actual as they are, do not differ in principle, at least not so as to discontinue the use of the term "persister" in the context of tumors as suggested [64]. Thus the use of the term in this work is continued.

Such persisters, by remaining in stasis they may be heavily coated, which both protects them and diminishes the rate of exchange with the environment, which is the main causeway to cytotoxic amenities, be they in the form of matter, as in chemotherapy, virotherapy, bacteriotherapy and immunotherapy [51, 65–67] or of energy, as in radiotherapy and thermotherapy [68, 69] while the extremely low metabolism diminishes further their susceptibility to such adverse affecting agents by not exposing the procedures and the respective effectors to their detrimental effect.

This alienation process, concluding in an auto-pathogenicity option, verges on speciation but, for whatever reason, has no genetic stability, at least not enough to define a novel genomic species; it may be due to a coherent strategy to develop variability to enhance survival, as the mutant phenotypes of unicellular organisms [43], or it may be an effect of the collapse of the established regulation network. Still, *per se* the alienation process, showcases the potential (or propensity) of a multicellular, single-genomed entity to create (or decline to) unicellular ones, fully alienated in attitude and function [43], despite any evolutionary genetic relations.

This fact may raise some questions on the phylogenesis of eukaryotic unicellulars, but, most intrinsically, of some viruses; especially, but not exclusively, of large size. They could be extra-lean, alienated constructs of reduced cellular agents, or escaping cellular debris [70, 71]. Especially DNA-genomed viruses, with their ability to form proviruses, integrate in host genomes and thus actually participate in the formation of new hologenomes within the host species [70], could well be regarded as such events that passed the test of Evolution and Natural Selection.

At a higher level, the cancer as a clinical entity could be perceived as a collective of tumors differing definitely in location/biocompartment and possibly in genetic content; they may be metastatic (and still highly unstable in genomic terms) but some of them might have occurred due to independent carcinogenic events [72], perhaps multiple spontaneous tumorigeneseis surging once the host immune system is overrun by the original tumor. In any case, even if very similar and metastatic, the different tumors are equivalent to different organisms, as they are separate and too distant to exchange matter. The current consensus seems to favor a concept of a surge of different tumors. The idea is functionally similar to a forest or a festering infectious disease. But the possibility that dispersed tumors are occasionally somehow communicating or even coordinating their attitudes should be explored henceforth, and it could be applicable to different foci of disseminated microbial infections [73, 74].

8.3.2 The Tribal Stage: The Colonial State

The colonial state [75], for want of a better name, refers to a conglomerate of cells, ostensibly of one unicellular organism (*conspecific*)—although this may be debated—that grow in the same environment and context and are directly interdependent, while coordinating and cooperating [76]. The latter prerogative is the most important. Cells—or morphological or functional groups of cells—that originate clonally from the same progenitor cell are called *ramets*, while the functional aggregation of ramets of the same origin (not only conspecific, but of the same pedigree line) define a *genet*, actually an aggregative organism [75]. It is important to denote that ramets are clonal *modules*, understood as individual cells or functionally integrated aggregates of cells [77], but all *modules* are not ramets. They may be zooids [78] or any of a set of microbial morphotypes, such as the cube of *Sarcina* [79] or the "bunch of grapes" clusters of *Staphylococcus* [80], which reminds post-zygote animal cell arrangement [81, 82]. The modules, of whatever name and nature/Kingdom of Life, may form colonial aggregations such as colonies, polyps, genets [75], or not. In the latter case they opt for solitary (occasionally unicellular) existence. On the other hand, it is generally understood that clonality is a prerequisite for coloniality [75, 83], but this may not be the case, as will be discussed later-the negative example being the mixed species ("multispecific") biofilms [84–86].

The concept of the intermediate, colonial state of organization may be defined as "cells not communicating through junctions but exchanging signals, so as to differentially share and divide tasks and coordinate responses and attitude so as to adapt". The *contact* principle of a previous definition applied to colonial animals [83] may be ill-advised, and the same goes with *clonality*, in the same definition. Regarding the former, the cells in biofilms may be lying at quite some distance—in relative terms—while in colonies there are limitations to the density of the cells, such as the density-dependent inhibition [46]. In the case of the colonial animals, the constitutive Zooids are kept apart, off-contact, due to their respective exoskeletons, which allow communication through pores [78].

Regarding the latter, the colonial state of organization falls between the community of unicellular organisms that *may* share information by biochemical, relatively long-range, space-agnostic, non-contact communication approaches on one hand, and, on the other, the differentiated, centrally controlled procedures of the multicellular state with cells in close physical proximity—actually in contact- and physical continuity. In the multicellular state of organization, the cells communicate only within a spatial barrier clearly defined and strictly controlled; so much so that *in extremis* they may be induced to undergo programmed cell death (*apoptosis*) in a most standardized, regulated, shipshape and optimized manner [87].

The colonial state does have spatial limitations as well: the ones of the colony, and occasionally a defined barrier in the form of some connective bioproducts. A biofilm, an exoskeleton or some other form of connective tissue more or less organized or canonical, defines them spatially and they have been understood earlier as an intermediate "between unicellular and multicellular life states" [88]. Occasionally within the connective substrate the cells are different in form than when independent, with the paradigm being the *sessile* forms of the cells embedded in a biofilm compared to the free-floating, *planktonic* forms [89].

A unicellular organism is capable of solitary existence [34] for its full circle repeatedly, meaning the offspring cells may do the same. These offspring cells though may settle very close to each other, and create a pedigree colony, in cases such as budding [90], where there is a distinction between maternal and daughter cell. This contrasts to the *per se* clonal concept of the simple division (*fission*) enacted by many bacteria [91] and fungi [92], but animals also [75], which results in two identical daughter cells from a single parent cell.

Differentiation occurs due to age and/or environmental challenges within pedigree, conspecific colonies. The switch phenotypes of unicellular fungi offer a proper example [93, 94]. Differentiation also occurs also in the case of fungal mycelia, which are multicellular lifeforms by definition [60], but in terms of cellular mechanics consist a small evolutionary step beyond the unicellular fungal states. Unicellular fungal states, such as yeasts, and mycelial ones may co-exist alternatively within the same species/strains depending on environmental conditions: the paradigm is the dimorphic fungi [95, 96]. Mycelia, similarly to unicellular yeast states, are produced by successive division mechanisms but, contrary to—most of—the latter, in a vectored sense: each cell does not divide uncontrollably, even under the best of conditions, but follows an ontogenetic pattern, the *apical growth* [97–99], which seems to abide to some quorum sensing principles, however crude [60, 100, 101]. including density-dependent inhibition [46]. Similar patterns of strict spatial limitations in the binary cell division are also the case in some bacteria, such as the genus *Sarcina* [79], thus underlining the existence of an organization and division pattern between uni- and multi-cellularity, where cells are not necessarily differentiated in terms of morphology, but are assigned different positions and possibly roles and fates [75].

Except fungi and bacteria, the colonial state is evident amongst animals: the *Zooid*, seen in bryozoans, is a single animal organism, partaking in a colonial state, such as the coral. Some exoskeleton or connective tissue connects the zooids of the full-size animal. The colonial state may be monomorphic or polymorphic; in the latter case, the zooids differ in structure and function. In bryozoan colonies the *autozooids* collect feed and supply nutrients to the heteromorphic zooids [78].

8.3.3 The Apperceived Apex: Multicellularity

The multicellular organism is characterized by the high-level physiological integration among its cells, which are in contact or at least in physical continuity by contact. More or less elaborate matter and energy exchanges take place among the constituent cells, the latter effected by electric microcurrents and chemical signals, and meant to allow the exchange of data and the transmission of commands [102].

The interest of the community/organism outpaces the one of the individual cells so much that the differentiation of aims to promote their role within the Whole and not to best adapt themselves within this Whole. The adaptation of the cells is functionalistic, not survivalistic and this is proven by the ability of the Whole, the organism, to issue by chemical or electric signals executive suicidal orders (apoptosis) [87, 103] for some expediency reason or simply to follow a given blueprint. This consideration of the individual cells as expendables is not found in the colonial level, where the cells aggregate and coordinate for mutually beneficial purposes. In the multicellular organism the cells are not subjects, but rather objects and the importance as defined by reproduction level and survival priority lies with the entire organism, which issues central, not collective, orders for adaptation to evolving situations and changing conditions. Thus a cell may be *replaced*, if damaged/necrosed, or *substituted*, by a different, differentiated one, after its apoptosis, for purely ontogenetic reasons; the transformation of the tissue and the organism. The differentiation may be restricted to biochemistry and physiology, which refer to its function, or it may include extensive differentiation that encompasses its form, so as to be optimized for its functionality and location, and also the concomitant (de)activation of specific genomic routines.

The multicellular nature is a high-end condition and not steady: multicellular organisms do have to transit through unicellularity within their life cycles, either to procreate, or to develop and maintain their body form and function [37]. Mammalia transform from unicellular gametic forms to multicellular ones, and dimorphic fungi from multicellular mycelia to unicellular states [95, 96]. The transformation includes colonial intermediates: The zygote after the initial rounds of divisions produces quorum-sensitive but still omni-potent daughter cells; if one of them is detached, it may produce a fully-fledged multicellular offspring, while it is promptly replaced at its original position by an *ad hoc* produced replacement cell; similar attitude is observed in the case of excised tissue from a mycelium, while the procreation routine is active [60]. Mycelia may transform to unicellular states either directly, by producing conidia, or indirectly, through some form of colonial transformation, including pseudomycelium.

Mycelial fungi are multicellular organisms, and they do present sometimes highly differentiated organs and occasionally tissues also [90]. The existence of coenocytic mycelia in lower fungal taxa, such as Zygomycetes [104], underlines the admission that non-cellular constructs, multinucleate and stemming from cellular aggregations that lost their individuality towards a community *are* multicellular organisms. This is also observed with Protozoa and slime molds, forming multinucleate, multicellular but syncytial *plasmodia* [105], and even with Metazoa, with the linear muscle fibers being multinucleate syncytial myotubes and a result of repeated cell divisions, following which the individuality of cells was compromised [106]. What is of interest is that contrary to older schemes, which considered multicellularity a prerogative of the eukaryotes, there are truly multicellular bacteria, as the Cyanobacteria [107] and the Actinobacteria [108], but also gigantic coenocytic bacterial organisms, multinucleate and macroscopic in size such as the *Thiomargarita* genus [109].

The strict, if not severe, regulation of the multicellular organism would suggest as a prerequisite the genomic and special homogeneity of the constituent cells; conspecificity at the very least, if not clonality/uninuclearity in all the cells of a multicellular organisms would be expected. Some room should be made for diversions as are the gametic cells in diploid organisms, the recombined populations of the B-cells that are hardwired clonally for one specific paratope (antigen-binding site) of their characteristic antibodies and the differential accumulation of mutations in different cells during the lifetime of an organism, especially in mutagenic environments, which produces *mosaicism* [110, 111].

But this is only part of the story: the *hybrid* organisms do not defy the paradigm, as their cells have all the same nucleus/genomic content, although this comes from two different germlines/genomes with different sequences or even chromosomes. The organism itself is uninucleate [112, 113]. And the same applies to transgenic organisms, where specific genomic loci are edited (inserted, excised, replaced) by various intervention techniques [114]. But the *chimerism* supports a paradigm shift in multicellularity, as some cells have different genomes than others in the same organism [110, 111]. The animals transplanted with xenologous grafts (grafts from different species) or even heterologous ones (from different individual of the same species) are perfectly functional multicellular organisms that operate with organs bearing different genomes (occasionally of interspecific variability), than that of the whole organism, while the same is true for grafted plants [115]. But the paradigm change is the concept of the holobiont and its accompanying hologenome [38, 40]. Once the mutualistic and commensalistic relationship of the native microflora of a macroorganism has been described as an integral part, if not an organ of a higher lifeform, the holobiont [116], genomic uniformity becomes obsolete and tolerance is afforded to genetic instability as well, as the microbiome changes during the lifetime of the host and according to the environmental conditions to no small an extent.

8.3.4 Meta-organisms: Functional Networking over Distance, Distributed Sensing, Coordination and Swarming

The second additional state of life as proposed herein lies beyond the multicellular organism. It is a direct counterpart of the colonial state of the unicellular organisms, transferred to multicellular-level ones, in that it tolerates physical distance amongst its constituents. What is of extreme importance is that it substitutes the need of close contact within a colony (the "how close" is an open issue) by the ability to communicate within networks that may differ in complexity and distance tolerance/range, especially—but not necessarily—in a contactless manner [117–119]. One legitimate question would be whether such aggregations, from forests to hives and swarms, may be considered *Meta-organisms* [36], similar to the Colonial formats, i.e. the pseudomycelia of the yeasts [97], the polyps of cnidarians [75, 120] and the different aggregations of bacteria, instead of being understood as a special, stricter social format.

A further question arises in whether the defining quality for a Meta-organism would be a hidden functional routine, embedded into its decision-making circuitry—not only in its DNA- that would partially transform the individual, solitary multicellular organisms to parts of the same Meta-organism. This transformation, so obvious in the alteration of physiology and ethology of locusts when creating a swarm, must- or could- be reversible. When conditions allow, or some counter-trigger is primed, the swarm members may revert to their solitary life patterns, attitude and routines [121, 122]. This is similar to what happens in a lower iteration of solitary and aggregative states, between colonial and purely unicellular forms: the sessile cells of a biofilm, which come from planktonic forms undergo transition under the control of some switch [89, 96, 123] and may revert back to sessile form, which differs in morphology and functions, under different conditions, including electrostimulation [124, 125].

The objective of pondering upon this idea is whether entities that are currently classified as organisms may be wholly integrated into a *Meta-organism* [36], which would equal some of the currently acknowledged social structures. Society succeeds biology when there is a spatial resolution of two autonomous, fully-fledged organisms *and* some autonomy of the individuals. When this distance does not exist but there are two autonomous organisms, there is the symbiosis phenomenon. The thing is: when the symbiont is unable to survive by itself, when it is an obligatory symbiont, then the symbiosis is a social or biological pattern? And the organism is the holobiont, or the symbiont(s) and the host are (two) separate individuals?

The issue becomes more perplexed if the integration of different species/genomes of the same order (for example, all of them symbionts) is taken into account. The multimicrobial biofilms [126, 127] are a perfect example. Another is set by trees and plants in general, which communicate chemically by airborne chemical messenger molecules, the Volatile Organic Compounds-VOC [118] and through the airwaves [119]. But among plants there is also a subterranean network of message/data exchange [117, 128]. As a result, there emerges another case that begs definition as the abovementioned two cases are different: through the air it is contactless communication, by sound, vibration or chemical moieties. The first two are similar to the vocal or other sound coding and perception of humans and animals and the third to the sense of odor; all of them understood as interindividual communication, and thus a societal, not physiological function. Subterranean communication may be chemical (by release and diffusion of chemical moieties), audio or anything else, as many waves are transmitted through the soil, perhaps electric ones too. But it may also be enacted through direct contact, as parts of the roots may be extending so as to touch the opposite numbers of other trees and thus create a network directly connecting the—or some—plants of a given site. Whether this contact should be understood as communication by contact code, like the different kinds of touch in animals and thigmotropism in general [129] or is reminiscent of the contact communication of neighboring cells in a tissue of an organism, is the focal issue that should be resolved.

The above concept may be trimmed down a bit to include only similar plants creating a conspesific network. In this, single-species iteration, things are a lot easier to digest than in the multi-species concept which verges a lot to the concepts of chimeric entities [130] if not of holobionts [38–40, 116]. The conspecific network in the biofilm iteration, which refers to unicellular constituents, may be considered a collective or a colony, and is similar to the connective tissues of evolved animal species whence individual cells are embedded in—and communicate through—extracellular matrix (ECM) produced by them [131, 132].

Of course the most intricate issue in this respect lies with the different animal species, the members of which are able to communicate interindividually with each other to form a hive [133, 134] or enact swarming [34, 122, 135]. The suicidal mode of swarms, packs and other groupings in some cases is just as inflexible as the apoptosis enacted in multicellular organisms. There are chilling social examples of orderly dispose of unfit individuals to spare the vitality/survivability of the group, or the optimal allocation of limited resources—and this without taking into consideration the need for conformation, which is enacted with exile— the lone-wolf paradigm. The Custom of Keos in the ancient Greek Island of Kea/Ceos (Strabo Geor X.5,6) is an example of autothanasia; the undertaker wolves in 1st millennium central Asia (Bactria and Sogdiana) [136] is pure senioeuthanasia [137] as is the ritualistic death of the sick Indian anchoret/gymnosophist Kalanus (Diod. 17.107) in fire. All are cases of seniocide, social behaviors of extreme cruelty—or determination [138], verging to the one of apoptosis, in different human societies. Similar are these of animals, as exemplified by the Last Trip of the elephants to their selected, perhaps in a sense consecrated, grounds to die [139].

8.3.5 The Code Dimension

From the above it is obvious that the new system considers the functional integration as the key element instead of the physical continuity/immediacy which defines the multicellular organism; it is true that the latter is a prerequisite, not a defining condition, as spatial immediacy does not automatically means a multicellular state, as already said (attached daughter budding cells, pseudomycelia). But under the new point of view, distance becomes somewhat irrelevant, and the physical division underlines the diminished consequence of distance: It is the concept of the Abstract Genome, whence the sum of the genetic loci and determinants of a species, understood in terms of length of coding nucleic acids' sequences but also other carriers of heredity (as are the centrosomes [140–143]) are not co-located within each of the vegetative bodies (cells or virions) but only conditionally so. The full set may be present in only some individuals (cells/organisms), or in transient forms during the ontogenesis, or even in no occasion. For example, the gonochoric life-forms [144, 145] already challenge the original concept, as two compatible and not identical members are needed for procreation and the entire genomic context, in qualitative terms (meaning type of chromosomes) was present to only part of the population, namely to the heterogametic sex. In mammals it is the male, in the XY sex determination system, in birds the female, in the WZ sex determination system [146].

Even more prominent such cases are the divided bacterial genomes [147] and the segmented viral genomes [148], especially in the case of multipartite viral genomes [149]. The latter imply that for a successful and productive viral infection, enacting the lytic cycle, the co-existence of carriers of the entire genome is indispensable. This, in turn, suggests nothing less than the need to collectively furnish the entirety of the viral genomic sequences in a spatiotemporal relevance to achieve the propagation of the virus [150].

The above considerations introduce the concept of the sub-genome: humans and other gonochoric organisms [144, 145], along with the viruses with segmented and multipartite genomes [148] constitute a class of a vertical division of the genome: all of the-possibly unequal- parts are needed to procreate. Although the multi-chromosome genomes are by definition fragmented, they follow very strict and specific rules that define the exact size, content and dosage of the different fragments for a standard propagation; deviations, shortages as much as overabundance (aneuploidy) create physiological problems to the probands. The multipartite genome implies that within any module, be that colonial, multicellular organism or community/swarm/Meta-organism, all parts of the genome have to be included by inclusion of their carrier particles, possibly to uneven quota. If the multipartitism refers to sexual chromosomes there may be an exception if asexual reproduction is achievable, irrespective of the state of the module (cell, colony, organism, Meta-organism/swarm/community).

However, there is another case, usually observed with the bacteria but this may be oversimplification: it is the existence of additional, detachable genomic elements, principally superfluous ones, such as plasmids and transposons, that may circulate laterally and supplement or exchange the genomic content [91, 147, 151, 152]. Such events happen not only in Bacteria, but within the other two domains of Life [153], and occasionally trans—domain such processes are reported [154]. The concept of some sequences inserted, to upgrade, or detached, to lighten a given genome is old but remains fascinating: it defies any notion of genomic identity within an aggregation of cells, mainly-and especially-within a clonal colony, once genes are inserted laterally from other similar (conjugative cells) or dissimilar (environment, viruses) donors.

The issue comes into prominence from the colonial state and upwards. Within a swarm or collective, expanded and unmodified genomes may be observed simultaneously. The obvious case is after a viral disease embeds viral DNA (of infectious or reverse-transcribed origin) plus transduced sequences into the genomes of a number but not all of the host's cells. At the multicellular state, where, generally speaking, much stricter rules apply, especially for homogeneity in genomic terms, there are the cases of chimeric organisms, where genomic homogeneity among the constitutive cells is compromised. They are defined as including cells (or tissues or organs) from two or more genetically distinct individuals, conspecific or not. Animals receiving transplantation or transfusion and plants subjected to inoculation may be included in the category [155].

The concept of the sub-genomes is boosted by the newer understanding of holobionts and hologenomes [40]. The idea that a macroorganism is a functional whole with its associated, appendix microbiomes (the plural denotes the rather determinative spatial separation amongst them, as in the gut microbiome and that of the armpit). Taken as an organism, the holobiont defies the idea of genomic homogeneity amongst the cells of a multicellular organism, it is a chimera and, what is much more important, it may be difficult to imagine a *proper* multicellular organism, at least a higher one, strictly monogenomic, devoid of functional interaction with microbiota. Germ-free organisms are plagued by a number of ailments as the microbiome performs some key functions, including an important part of digestion [156]. In the communal state things are much more complicated, as even genomically identical macroorganisms, such as animals produced asexually, may harbor microbiomes of vast variability but extreme importance. The holobiont is occasionally called "super-organism" [157], causing some confusion; it consists a step up from the multicellular organism, which adds additional and much different cells, but not a transition to the higher state of life described herein as Meta-organism.

Last comes the case of genomic heterogeneity within colonial organisms. It is obvious that in the colonial context genomic homogeneity would be even less of a priority. The polymicrobial (or multispecies, or mixed-species) biofilms [47, 158] are ample proof, as they may contain sessile cells of extremely distant taxa, occasionally spanning across all three Domains of Life [84–86]. Within the biofilm cells may not be in direct contact, but they communicate physically with each other through the extracellular matrix into which they all thrive and contribute, which sums up the notion of a colony pretty well.

8.4 Guided Self-Positioning and Regulation. A Prime Window for Disrupting Disease?

The concept of Quorum Sensing -QS explains how individual unicellular organisms, especially bacteria, can be optimally arranged in spatiotemporal terms so as to survive, get established and then thrive under challenge in hostile environments [159] such as the one created inside an infected multicellular organism by its immune system-if properly functional. The idea of disrupting QS by deploying Quorum Sensing Inhibitors -QSI or Quorum Quenchers -QQ is obvious [160–162]. Formation is an enhancer in physical contest as it improves different aspects of the confrontation mechanics [163]. QS might be the reason for a non-growing part of a mycelium, if transferred to a fresh, benign spot to start growing, while the emptied spot is not covered by resumed growth of neighboring hyphae [60]. The QS modalities, should these be the causative agents, obviously remain in place and inhibit the growth even after the biomass that produced them has been removed. So disrupting this spatial arrangement may reduce the virulence and growth potential of the unicellular agent. A notion of QS allows the heterologous bilayer formation (one layer of cells on another of basic membrane), so typical of animal epithelia [202] and its cancellation is amongst the hallmarks of carcinogenesis [56].

But the *multicellularization*, and the formation of a higher -level organism in general, as proposed herein, is much more than arrangement in space and group forming: the latter is the hallmark of society and in systems science a very low integration setup. The multi-assimilation of constituents to form a new organism includes the permanent or semi-permanent but predictable assignment of diverse if not contradictory duties for specialization by roles and functions and the extreme altruistic options of suicide of individuals for the profit of the Whole, for the Greater Good in terms of survival and fitness [37]. In this prerogative lies the capital interest of how a group will be apperceived. If it is an organism, then there is a central, or at least a structured command system, as it may be distributed for reasons of ergonomics, as in multi-brain animals [164]. Commands needed for survival through adaptation, for procreation and homeostasis are informed by sensory input and communicated somehow. If this communication, or rather if this open-loop activities are shut down or hacked, then the construct will collapse.

In a QQ-caused incident the bacteria would lose formation and coordination, and their concerted action would be disrupted, becoming suboptimal and their attack less cohesive and effective. The reaction from the immune system and any possible ongoing intervention would have a larger temporal window into which to act so as to contain the infection and much better chances to defeat and clear it, as the numerous but randomly dispersed cells might be more exposed to cytotoxic actions of any kind, phagocytosis included.

On the other hand, hacking of the command and control system (sensory items and pipelines, perception and decision circuitry and executive pipelines, motor ones included) may resolve the infection almost on the spot. Depending on the exact application and the bacteria/pathogen cell type, the immediate effects on the invading cells, especially but not necessarily prokaryotic, are expected to disrupt their communications and thus cause disorganization which may result in any of the following conditions: the switching off of exotoxin production and thus relief of the invaded host of the toxic effects, which may well be immunosuppressing; the inhibition of the translation and excretion of enzymes that modify the environment to the benefit of the invading cells; the arrest of their propagation; the suppression of adaptational processes, including but not limited to the formation of endospores [165] and protoplasts/spheroplasts [166]; the enactment of conjugation, transposition and horizontal/lateral gene proliferation routine [91, 152, 167, 168]; the initialization of bacterial apoptosis [103] and, possibly, aggression of any degree among cells or group of cells in an increasingly hostile surrounding to which they cannot effectively respond nor adapt. All the above assist the immune system, or any intervention, or both, being more effective in combatting the infection and thus In resolving it relatively fast and with minimal resistance, which in turn leads to lower and fewer fever outbursts, less accumulation of puss and less cascades of disruptive events such as cytokine storms [169] or rheumatic syndromes such as rheumatic fever [170, 171]. To the above, one may add decreased direct tissue damage, expenditure of energy and material resources of the host, potential for communication of the infection to other hosts (Transmission) or to other biocompartments of the same host (Dissemination) and toxic adverse effects of administered drugs [172]. Whether the proliferation of drug resistance within and among pathogenic populations, especially through Lateral Gene Transfer -LGT [91, 152], may also be included in the list of the interceptable events through the hacking of the collective decision functions remains to be determined.

Similar if not identical effects may be brought upon tumors. Tumors are an example of organized anarchy; their cells have escaped the centralized regulation, especially the enactment of apoptosis and the inhibition of propagation, but show a number of attributes that suit organized, not fully unregulated groups: one is the existence and functionality of the persister cells. The reaction to their destruction is not clear. They may be replaced by another *vegetative* tumor cell turning persister/sleeper and going to hybernation, or a persister cell must be created *de novo* by transforming a benign tumor cell or even a normal one by some short of corruption. In the last case the best option for quick corruption to persister status may be the normal cells usually engulfed within any tumor.

Thus, if the concepts just overviewed are translatable to cancer treatment and get transferred from the infection control medical community, it *is* as if the medicinal Holy Grail has been recovered. It is much more than being able to use the same drugs. This is doubtful as they target widely different physiologies, the antibiotics and the antineoplastics. It is the similarity in concept which may allow similar procedures and protocols, similar diagnostics and management options and, in the end, similar amenities, (bio)chemical or not, but targeting common or similar functional entities: If a change of the pH disrupts the chemical signaling, then hacking the pathobiont cells into effecting themselves this change in their vicinity would disrupt the networking and their command and control function.

8.5 The Magic Wand in Biosecirity and Health Promotion

If there is merit in the suggestion for the four states of life concept, it is likely to prime both medium and long-term developments. Regarding the former, the focus will be in discovering or developing amenities that would work disruptively to one or more state(s) of life but less so to others, so as to combat microbial diseases in different species as a fulfilment to the One Health principle [173–175]. This should occur without directly damaging neither the harmful, pathogenic cells proper, an approach that would curtail toxic effects by endotoxins [176, 177] and hypersensitivity reactions [178]; nor the host and its symbionts, which is the most important attribute for a therapeutic amenity. It would be an issue of Artificial Intelligence to tap the way this approach would fit into agnostic treatment. The latter is understood as a targeted, informed treatment *without knowing* the pathogenic species (or even higher taxonomic grades of the pathogen), as agnostic surveillance, diagnostics and intervention seem to be setting the vogue [179–183] because of the potentially indefinite numbers of possible pathogens, due to spontaneous emergence or deliberate engineering [184–188].

Similar is the case with cancer therapy, but the focal issue is the prospect of amenities applicable in both cases. It will not be important how different the cancer cells are compared to the bacterial or other microbial ones, nor their presenting different immune markers and targetable molecules. The importance is bound to shift to how similar they are in status, operational routines and networks and procedures [43, 60] and thus how common disruptive solutions are likely to unsettle both; this is squarely in the realm of Integrative Biology, although in a rather peculiar manner [189, 190]. Electrical modalities come readily to mind [191–193] but other principles should be investigated.

Regarding the latter, if indeed the integration level can be regarded as something more fluid and less inflexible than it currently is; if distance without physical continuity is not an absolute obstacle in considering two elements as of the same organism due to their level of integration and share interests and command and control functions, then there are some very important possibilities exceeding the obvious ones in cancer and microbial diseases. Such a fact, if proven, would inadvertently mean the existence of communication amenities and codes hitherto unidentified, unbroken or highly underestimated; to understand and identify them may be a revolution similar or greater than the RNA Revolution [194], which completely recast our understanding of the regulatory functions, the mulltifunctionality of RNAs and the organization of the genome; perhaps even the definition of the gene, as a *transcribable sequence* rather than a *translatable* one could be more fitting.

There is no way to know if this augmented integration occurs with alternative functions of existing biochemical entities (molecules), or with other, known or totally unknown ones. For the latter one may speculate; but for the former, such an intercellular transmission function is conceivable for some short of small RNAs. Different RNAs might be found to be used for transmitting standard, highly important, actually vital messages among cells in direct contact or across short distances, within a protected environment, i.e. a biofilm [195] or an extracellular matrix in connective tissue [131, 132] and with extreme consistency: the messages might be found to be drafted and conserved within the nucleus, in DNA sequence—possibly the so called "junk DNA". One geneticist's Junk may be another's Nobel Prize. In this respect, with a much more complicated and multi-aspect understanding of intercellular relationships and their stratification, it might be possible to identify methods to tackle other ailments as well. And this intervention would be direct, not through effects borne by microbiomic manipulation and management [196–198]. For example, ailments due to deterioration/dysfunction may be amenable to management by manipulation of a new, hitherto undetected communication function that enables unknown interactions, similar to the fairly recently identified importance of the gut-brain axis—and the prospects for therapeutic interventions targeted on it [199–201].

References

1. Desnues C, Boyer M, Raoult D (2012) Sputnik, a virophage infecting the viral domain of life. Adv Virus Res 63–89. https://doi.org/10.1016/B978-0-12-394621-8.00013-3
2. Koonin EV, Wolf YI (2012) Evolution of microbes and viruses: a paradigm shift in evolutionary biology? Front Cell Infect Microbiol 2:119. https://doi.org/10.3389/fcimb.2012.00119
3. Raoult D, Forterre P (2008) Redefining viruses: lessons from Mimivirus. Nat Rev Microbiol 6:315–319. https://doi.org/10.1038/nrmicro1858
4. Wogan GN, Hecht SS, Felton JS, Conney AH, Loeb LA (2004) Environmental and chemical carcinogenesis. Semin Cancer Biol 14:473–486. https://doi.org/10.1016/j.semcancer.2004.06.010
5. D'orazio J, Jarrett S, Amaro-Ortiz A, Scott T (2013) UV radiation and the skin. Int J Mol Sci 14:12222–12248 [Internet]. https://doi.org/10.3390/ijms140612222
6. Carpenter DO (2013) Human disease resulting from exposure to electromagnetic fields. Rev Environ Health 28:159–172. https://doi.org/10.1515/reveh-2013-0016
7. Dianzani F (1975) Viral interference and interferon. Ric Clin Lab 5:196–213 [Internet]. https://doi.org/10.1007/BF02908284. Cited 18 Aug 2022
8. Teshome S (2012) CIA targeted assassinations by induced heart attack and cancer [Internet]. ethiothinkthank. https://ethiothinkthank.wordpress.com/2015/12/06/cia-targeted-assassinations-by-induced-heart-attack-and-cancer/
9. Moghaddar M, Radman R, Macreadie I (2021) Severity, pathogenicity and transmissibility of delta and lambda variants of SARS-CoV-2, toxicity of spike protein and possibilities for future prevention of COVID-19. Microorganisms. https://doi.org/10.3390/microorganisms9102167
10. Furuta Y, Cheng C, Zorigt T, Paudel A, Izumi S, Tsujinouchi M et al (2021) Direct regulons of AtxA, the master virulence regulator of *Bacillus anthracis*. mSystems 6:e0029121. https://doi.org/10.1128/mSystems.00291-21
11. Taylor-Joyce G, Manoharan S, Brooker TA, Hernández-Rodríguez CS, Baillie L, Oyston PCF et al (2023) The influence of extrachromosomal elements in the anthrax "cross-over" strain *Bacillus cereus* G9241. Front Microbiol 14:1113642. https://doi.org/10.3389/fmicb.2023.1113642
12. Oliveira ASF, Ibarra AA, Bermudez I, Casalino L, Gaieb Z, Shoemark DK et al (2021) A potential interaction between the SARS-CoV-2 spike protein and nicotinic acetylcholine receptors. Biophys J 120:983–993 [Internet]. https://doi.org/10.1016/j.bpj.2021.01.037

13. Russo M, Sogari A, Bardelli A (2021) Adaptive evolution: how bacteria and cancer cells survive stressful conditions and drug treatment. Cancer Discov 11:1886–1895. https://doi.org/10.1158/2159-8290.CD-20-1588
14. Pouysségur J, Marchiq I, Parks SK, Durivault J, Ždralević M, Vucetic M (2022) 'Warburg effect' controls tumor growth, bacterial, viral infections and immunity—genetic deconstruction and therapeutic perspectives. Semin Cancer Biol 86:334–346. https://doi.org/10.1016/J.SEMCANCER.2022.07.004
15. Norton L, Massagué J (2006) Is cancer a disease of self-seeding? Nat Med 12:875–878 [Internet]. https://doi.org/10.1038/nm0806-875
16. Hernández G (2022) Schrödinger and the possible existence of different types of life [Internet]. Front Microbiol. https://www.frontiersin.org/articles/10.3389/fmicb.2022.902212
17. Yolles M, Frieden R (2022) Viruses as living systems—a metacybernetic view. Systems. https://doi.org/10.3390/systems10030070
18. Pearson H (2008) "Virophage" suggests viruses are alive. Nature 454:677 [Internet]. Nature Publishing Group. https://doi.org/10.1038/454677a. Cited 22 Feb 2021
19. Ding B (2010) Viroids: self-replicating, mobile, and fast-evolving noncoding regulatory RNAs. Wiley Interdiscip Rev RNA 1:362–375 [Internet]. https://doi.org/10.1002/wrna.22. Cited 3 Nov 2020
20. Hyde ST (2015) Crystals: animal, vegetable or mineral? Interface Focus 5 [Internet]. https://doi.org/10.1098/RSFS.2015.0027. Cited 28 Feb 2023
21. Palacci J, Sacanna S, Steinberg AP, Pine DJ, Chaikin PM (2013) Living crystals of light-activated colloidal surfers. Science (80-) 339:936–940 [Internet]. American Association for the Advancement of Science. https://doi.org/10.1126/science.1230020
22. Hallsworth JE, Mancinelli RL, Conley CA, Dallas TD, Rinaldi T, Davila AF et al (2021) Astrobiology of life on earth. Environ Microbiol 23:3335-3344 [Internet]. https://doi.org/10.1111/1462-2920.15499. Cited 29 Nov 2022
23. Davies PCW, Lineweaver CH (2005) Finding a second sample of life on earth. Astrobiology 5:154–163 [Internet]. https://doi.org/10.1089/AST.2005.5.154. Cited 11 Apr 2022
24. Cleland CE, Copley SD (2005) The possibility of alternative microbial life on Earth. Int J Astrobiol 4:165–173. https://doi.org/10.1017/S147355040500279X
25. Seager S, Bains W (2024) The search for signs of life on exoplanets at the interface of chemistry and planetary science. Sci Adv 1:e1500047 [Internet]. American Association for the Advancement of Science. https://doi.org/10.1126/sciadv.1500047
26. Diwo C, Budisa N (2018) Alternative biochemistries for alien life: basic concepts and requirements for the design of a robust biocontainment system in genetic isolation. Genes (Basel) 10 [Internet]. MDPI AG. https://doi.org/10.3390/genes10010017. Cited 21 Sep 2020
27. Tetz VV, Tetz GV (2020) A new biological definition of life. Biomol Concepts 11:1–6 [Internet]. NLM (Medline). https://doi.org/10.1515/bmc-2020-0001. Cited 16 Nov 2021
28. Kambouris M (2021) The concept of humanome and the microbiomic dimension. In: Kambouris M (ed) Genomics in biosecurity, 1st edn. Elsevier Academic Press, London, pp 15–28
29. Dong Q, Xing X (2018) Cancer cells arise from bacteria. Cancer Cell Int 18:205 [Internet]. https://doi.org/10.1186/s12935-018-0699-4
30. Dart A (2020) Bacterial residents. Nat Rev Cancer 20:413 [Internet]. https://doi.org/10.1038/s41568-020-0284-8
31. Nejman D, Livyatan I, Fuks G, Gavert N, Zwang Y, Geller LT et al (2020) The human tumor microbiome is composed of tumor type-specific intracellular bacteria. Science 368:973–980. https://doi.org/10.1126/science.aay9189
32. Schorr L, Mathies M, Elinav E, Puschhof J (2023) Intracellular bacteria in cancer—prospects and debates. npj Biofilms Microbiomes 9:76 [Internet]. https://doi.org/10.1038/s41522-023-00446-9
33. Sumpter DJT (2006) The principles of collective animal behaviour. Philos Trans R Soc Lond Ser B, Biol Sci 361:5–22. https://doi.org/10.1098/rstb.2005.1733

34. Keya J, Suzuki R, Kabir A, Inoue D, Asanuma H, Sada K et al (2018) DNA-assisted swarm control in a biomolecular motor system. Nat Commun 9 [Internet]. https://doi.org/10.1038/S41467-017-02778-5. Cited 29 Sep 2021
35. De Souza DA, Kaftanoglu O, De Jong D, Page Jr RE, Amdam GV et al (2018) Differences in the morphology, physiology and gene expression of honey bee queens and workers reared in vitro versus in situ. Biol Open 7 [Internet]. Company of Biologists. https://doi.org/10.1242/BIO.036616. Cited 29 Sep 2021
36. O'Malley MA (2013) Metaorganism. In: Dubitzky W, Wolkenhauer O, Cho K-H, Yokota H (eds) Encyclopedia of systems biology. Springer New York, New York, NY, pp 1297–1299. https://doi.org/10.1007/978-1-4419-9863-7_75
37. Michod RE (2007) Evolution of individuality during the transition from unicellular to multicellular life. Proc Natl Acad Sci U S A 104(Suppl):8613–8618. https://doi.org/10.1073/pnas.0701489104
38. Gilbert SF, Sapp J, Tauber AI (2012) A symbiotic view of life: we have never been individuals. Q Rev Biol 87:325–341. https://doi.org/10.1086/668166
39. Richardson LA (2017) Evolving as a holobiont. PLoS Biol 15 [Internet]. https://doi.org/10.1371/JOURNAL.PBIO.2002168. Cited 11 Apr 2022
40. Morris JJ (2018) What is the hologenome concept of evolution? F1000Research, England, p 7. https://doi.org/10.12688/f1000research.14385.1
41. Berg G, Rybakova D, Fischer D, Cernava T, Vergès MCC, Charles T et al (2020) Microbiome definition re-visited: old concepts and new challenges. Microbiome 8:103. BioMed Central. https://doi.org/10.1186/s40168-020-00875-0
42. Cristescu ME (2019) The concept of genome after one century of usage. Genome 62:III–V [Internet]. NRC Research Press. https://doi.org/10.1139/gen-2019-0129
43. Trigos AS, Pearson RB, Papenfuss AT, Goode DL (2018) How the evolution of multicellularity set the stage for cancer. Br J Cancer 118:145–152. https://doi.org/10.1038/bjc.2017.398
44. Burrell RA, McGranahan N, Bartek J, Swanton C (2013) The causes and consequences of genetic heterogeneity in cancer evolution. Nature 501:338–345 [Internet]. Nature Publishing Group. https://doi.org/10.1038/nature12625. Cited 18 Aug 2022
45. Bacchus P, Nissen K, Berg J, Bråve A, Gyll J, Larsson C et al (2021) Civil-military collaboration to facilitate rapid deployment of a mobile laboratory in early response to COVID-19: a high-readiness exercise. Health Secur 19:488–497. https://doi.org/10.1089/HS.2021.0011
46. Cooper G (2000) The development and causes of cancer [Internet], 2nd edn. Cell A Mol. Approach. Sinauer Associates, Sunderland (MA). https://www.ncbi.nlm.nih.gov/books/NBK9963/
47. Gabrilska RA, Rumbaugh KP (2015) Biofilm models of polymicrobial infection. Future Microbiol 10:1997–2015 [Internet]. Future Science Group. https://doi.org/10.2217/fmb.15.109. Cited 8 Jul 2019
48. Venitt S (1996) Mechanisms of spontaneous human cancers. Environ Health Perspect 104(Suppl):633–637. https://doi.org/10.1289/ehp.96104s3633
49. Cadwell K (2015) The virome in host health and disease [Internet]. Immunity 805–813. Cell Press. https://doi.org/10.1016/j.immuni.2015.05.003. Cited 30 Sep 2020
50. Yusuf K, Sampath V, Umar S (2023) Bacterial infections and cancer: exploring this association and its implications for cancer patients. Int J Mol Sci 24. https://doi.org/10.3390/ijms24043110
51. Mughal MJ, Kwok HF (2022) Multidimensional role of bacteria in cancer: mechanisms insight, diagnostic, preventive and therapeutic potential. Semin Cancer Biol 86:1026–1044 [Internet]. https://doi.org/10.1016/j.semcancer.2021.06.011
52. Travaglione S, Carlini F, Maroccia Z, Fabbri A (2024) Special Issue "Bacterial Toxins and Cancer." Int J Mol Sci. https://doi.org/10.3390/ijms25042128
53. Mao Q, Jiang F, Yin R, Wang J, Xia W, Dong G et al (2018) Interplay between the lung microbiome and lung cancer. Cancer Lett 415:40–48 [Internet]. https://doi.org/10.1016/j.canlet.2017.11.036. Cited 5 Jul 2019

54. Gonzalez H, Hagerling C, Werb Z (2018) Roles of the immune system in cancer: from tumor initiation to metastatic progression. Genes Dev 32:1267–1284. https://doi.org/10.1101/gad.314617.118
55. Xia L, Oyang L, Lin J, Tan S, Han Y, Wu N et al (2021) The cancer metabolic reprogramming and immune response. Mol Cancer 20:28 [Internet]. https://doi.org/10.1186/s12943-021-01316-8
56. Hanahan D, Weinberg RA (2011) Hallmarks of cancer: the next generation. Cell 144:646–674. https://doi.org/10.1016/j.cell.2011.02.013
57. Buzby GP, Mullen JL, Stein TP, Miller EE, Hobbs CL, Rosato EF (1980) Host-tumor interaction and nutrient supply. Cancer 45:2940–2948. https://doi.org/10.1002/1097-0142(19800615)45:12%3c2940::aid-cncr2820451208%3e3.0.co;2-p
58. Liberti MV, Locasale JW (2016) The warburg effect: how does it benefit cancer cells? Trends Biochem Sci 41:211–218. https://doi.org/10.1016/j.tibs.2015.12.001
59. Biernaskie JM, West SA (2015) Cooperation, clumping and the evolution of multicellularity. Proc Biol Sci 282:20151075. https://doi.org/10.1098/rspb.2015.1075
60. Kambouris ME, Goudoudaki S, Kritikou S, Milioni A, Karamperis K, Giavasis I et al (2022) Beyond the microbiome: germ-ganism? An integrative idea for microbial existence, organization, growth, pathogenicity, and therapeutics. OMICS 26:204–217 [Internet]. https://doi.org/10.1089/OMI.2022.0015. Cited 27 Apr 2022
61. Vallette FM, Olivier C, Lézot F, Oliver L, Cochonneau D, Lalier L et al (2019) Dormant, quiescent, tolerant and persister cells: four synonyms for the same target in cancer. Biochem Pharmacol 162:169–176. https://doi.org/10.1016/j.bcp.2018.11.004
62. Sulaiman JE, Lam H (2021) Evolution of bacterial tolerance under antibiotic treatment and its implications on the development of resistance. Front Microbiol 12:431. https://doi.org/10.3389/FMICB.2021.617412/BIBTEX
63. Pirofski LA, Casadevall A (2020) The state of latency in microbial pathogenesis [Internet]. J Clin Invest 4525–4531. American Society for Clinical Investigation. https://doi.org/10.1172/JCI136221. Cited 25 Sep 2022
64. Decollogny M, Rottenberg S (2024) Persisting cancer cells are different from bacterial persisters. Trends Cancer 10:393–406 [Internet]. https://doi.org/10.1016/j.trecan.2024.02.002
65. Alberti C (2017) Prostate cancer immunotherapy, particularly in combination with androgen deprivation or radiation treatment. Customized pharmacogenomic approaches to overcome immunotherapy cancer resistance. G Chir 37:225–235 [Internet]. https://doi.org/10.11138/GCHIR/2016.37.5.225. Cited 27 Sep 2021
66. Nettelbeck DM, Leber MF, Altomonte J, Angelova A, Beil J, Berchtold S et al (2021) Virotherapy in Germany—recent activities in virus engineering, preclinical development, and clinical studies. Viruses 13 [Internet]. Multidisciplinary Digital Publishing Institute (MDPI). https://doi.org/10.3390/V13081420. Cited 20 Aug 2022
67. Aipire A, Mahabati M, Cai S, Wei X, Yuan P, Aimaier A et al. The immunostimulatory activity of polysaccharides from *Glycyrrhiza uralensis*. PeerJ 8 [Internet]. https://doi.org/10.7717/PEERJ.8294. Cited 20 Apr 2022
68. Johnson K, Chang-Claude J, Critchley A, Kyriacou C, Lavers S, Rattay T et al (2019) Genetic variants predict optimal timing of radiotherapy to reduce side-effects in breast cancer patients. Clin Oncol (R Coll Radiol) 31:9–16 [Internet]. https://doi.org/10.1016/J.CLON.2018.10.001. Cited 27 Sep 2022
69. Hirsch LR, Stafford RJ, Bankson JA, Sershen SR, Rivera B, Price RE et al (2003) Nanoshell-mediated near-infrared thermal therapy of tumors under magnetic resonance guidance. Proc Natl Acad Sci U S A 100:13549–13554 [Internet]. https://doi.org/10.1073/PNAS.2232479100. Cited 5 Dec 2023
70. Nasir A, Romero-Severson E, Claverie J-M (2020) Investigating the concept and origin of viruses. Trends Microbiol 28:959–967. https://doi.org/10.1016/j.tim.2020.08.003
71. Forterre P (2006) The origin of viruses and their possible roles in major evolutionary transitions. Virus Res 117:5–16. https://doi.org/10.1016/J.VIRUSRES.2006.01.010

72. Kim ES, Khuri FR, Hong WK (2002) Chemoprevention trials. In: Bertino JRBT-E of C (Second E, editor). Academic Press, New York, pp 457–472. https://doi.org/10.1016/B0-12-227555-1/00054-X
73. Melcher GP, McGough DA, Fothergill AW, Norris C, Rinaldi MG (1993) Disseminated hyalohyphomycosis caused by a novel human pathogen, *Fusarium napiforme*. J Clin Microbiol 31
74. Ribes JA, Vanover-Sams CL, Baker DJ (2000) Zygomycetes in human disease. Clin Microbiol Rev 13:236–301 [Internet]. https://doi.org/10.1128/CMR.13.2.236. Cited 20 Apr 2022
75. Hiebert LS, Simpson C, Tiozzo S. Coloniality, clonality, and modularity in animals: the elephant in the room. J Exp Zool Part B Mol Dev Evol 336:198–211 [Internet]. https://doi.org/10.1002/jez.b.22944
76. Jackson JBC (1977) Competition on marine hard substrata: the adaptive significance of solitary and colonial strategies. Am Nat 111:743–767 [Internet]. https://doi.org/10.1086/283203
77. Harper J (1980) Plant demography and ecological theory. Oikos 35:244–253
78. Shunatova N, Denisova S, Shchenkov S, Filippov A (2023) Colonial system of integration and communication pores in a polymorphic bryozoan *Dendrobeania fruticosa* (Bryozoa: Cheilostomata). J Morphol 284:e21601. https://doi.org/10.1002/jmor.21601
79. Canale-Parola E, Wolfe RS (1960) Studies on *Sarcina ventriculi*. J Bacteriol 79:857–862 [Internet]. American Society for Microbiology. https://doi.org/10.1128/jb.79.6.857-859.1960. Cited 22 Apr 2022
80. Turner RD, Ratcliffe EC, Wheeler R, Golestanian R, Hobbs JK, Foster SJ (2010) Peptidoglycan architecture can specify division planes in Staphylococcus aureus. Nat Commun 1:26 [Internet]. https://doi.org/10.1038/ncomms1025
81. Ajduk A, Zernicka-Goetz M (2016) Polarity and cell division orientation in the cleavage embryo: from worm to human. Mol Hum Reprod 22:691–703. https://doi.org/10.1093/molehr/gav068
82. Kumar M, Pushpa K, Mylavarapu SVS (2015) Splitting the cell, building the organism: mechanisms of cell division in metazoan embryos. IUBMB Life 67:575–587 [Internet]. https://doi.org/10.1002/iub.1404
83. Boardman R, Cheetham A, Oliver W (1973) Animal colonies. Development and function through time. Dowden, Hutchinson & Ross, Inc., Stroudsburg, PA
84. Ghannoum M (2016) Cooperative evolutionary strategy between the bacteriome and mycobiome. MBio 7:10–13. https://doi.org/10.1128/mBio.01951-16
85. Probst AJ, Birarda G, Holman H-YN, DeSantis TZ, Wanner G, Andersen GL et al (2014). Coupling genetic and chemical microbiome profiling reveals heterogeneity of archaeome and bacteriome in subsurface biofilms that are dominated by the same archaeal species. PLoS ONE 9:e99801 [Internet]. https://doi.org/10.1371/journal.pone.0099801. Cited 3 Jun 2019
86. Wargo MJ, Hogan DA (2006) Fungal-bacterial interactions: a mixed bag of mingling microbes. Curr Opin Microbiol 9:359–364. https://doi.org/10.1016/j.mib.2006.06.001
87. Elmore S (2007) Apoptosis: a review of programmed cell death. Toxicol Pathol 35:495–516. https://doi.org/10.1080/01926230701320337
88. Erez A, Kolodkin-Gal I (2017) From prokaryotes to cancer: glutamine flux in multicellular units. Trends Endocrinol Metab 28:637–644 [Internet]. https://doi.org/10.1016/J.TEM.2017.05.007. Cited 20 Oct 2021
89. Rollet C, Gal L, Guzzo J (2009) Biofilm-detached cells, a transition from a sessile to a planktonic phenotype: a comparative study of adhesion and physiological characteristics in *Pseudomonas aeruginosa*. FEMS Microbiol Lett 290:135–142 [Internet]. https://doi.org/10.1111/j.1574-6968.2008.01415.x. Cited 11 Feb 2019
90. Quijada L, Matočec N, Kušan I, Tanney JB, Johnston PR, Mešić A et al (2022) Apothecial ancestry, evolution, and re-evolution in thelebolales (Leotiomycetes, Fungi). Biology (Basel) 11. https://doi.org/10.3390/biology11040583
91. Gillings MR (2017) Lateral gene transfer, bacterial genome evolution, and the anthropocene. Ann N Y Acad Sci 1389:20–36. https://doi.org/10.1111/nyas.13213

92. Osumi M (2012) Visualization of yeast cells by electron microscopy. J Electron Microsc (Tokyo) 61:343–365 [Internet]. https://doi.org/10.1093/jmicro/dfs082. Cited 22 Feb 2021
93. Fries BC, Goldman DL, Casadevall A (2002) Phenotypic switching in *Cryptococcus neoformans*. Microbes Infect 4:1345–1352. https://doi.org/10.1016/S1286-4579(02)00006-0
94. Velegraki A (1995) In vitro susceptibility to itraconazole and fluconazole of switch phenotypes of *Candida albicans* serotypes A and B, isolated from immunocompromised hosts—PubMed. J Med Vet Mycol 33:83–85 [Internet]. https://pubmed.ncbi.nlm.nih.gov/7650586/. Accessed 22 Nov 2021
95. Nigg M, Bernier L (2016) From yeast to hypha: defining transcriptomic signatures of the morphological switch in the dimorphic fungal pathogen *Ophiostoma novoulmi*. BMC Genomics 17. BioMed Central Ltd. https://doi.org/10.1186/S12864-016-3251-8
96. Nadal M, García-Pedrajas M, Gold S (2008) Dimorphism in fungal plant pathogens. FEMS Microbiol Lett 284:127–134 [Internet]. https://doi.org/10.1111/J.1574-6968.2008.01173.X. Cited 21 Sep 2021
97. Gow NAR, Gooday GW (1982) Growth kinetics and morphology of colonies of the filamentous form of *Candida albicans*. J Gen Microbiol 128:2187–2194 [Internet]. https://doi.org/10.1099/00221287-128-9-2187. Cited 15 Nov 2021
98. Yang H, King R, Reichl U, Gilles ED (1992) Mathematical model for apical growth, septation, and branching of mycelial microorganisms. Biotechnol Bioeng 39:49–58 [Internet]. https://doi.org/10.1002/BIT.260390109. Cited 15 Nov 2021
99. Semighini CP, Harris SD (2008) Regulation of apical dominance in *Aspergillus nidulans* hyphae by reactive oxygen species. Genetics 179:1919–1932 [Internet]. https://doi.org/10.1534/GENETICS.108.089318. Cited 9 Nov 2021
100. Costa-Orlandi CB, Sardi JCO, Pitangui NS, de Oliveira HC, Scorzoni L, Galeane MC et al (2017) Fungal biofilms and polymicrobial diseases. J Fungi (Basel, Switzerland) 3:22 [Internet]. https://doi.org/10.3390/jof3020022. Cited 11 Feb 2019
101. Bonner JT (2004) Perspective: the size-complexity rule. Evolution 58:1883–1890 [Internet]. https://doi.org/10.1111/J.0014-3820.2004.TB00476.X. Cited 20 Apr 2022
102. Kolimechkov S, Seijo M, Swaine I, Thirkell J, Colado JC, Naclerio F (2023) Physiological effects of microcurrent and its application for maximising acute responses and chronic adaptations to exercise. Eur J Appl Physiol 123:451–465 [Internet]. https://doi.org/10.1007/s00421-022-05097-w
103. Hammer ND (2018) Microbial programmed cell death. Apoptosis beyond many ways cells die [Internet]. Wiley, pp 49–70. https://doi.org/10.1002/9781119432463.CH3. Cited 15 Aug 2022
104. Mehta BJ, Cerdá-Olmedo E (2001) Intersexual partial diploids of phycomyces. Genetics 158:635–641. https://doi.org/10.1093/genetics/158.2.635
105. Gerber T, Loureiro C, Schramma N, Chen S, Jain A, Weber A et al (2022) Spatial transcriptomic and single-nucleus analysis reveals heterogeneity in a gigantic single-celled syncytium. eLife 11. https://doi.org/10.7554/eLife.69745
106. Nadkarni AV, Heald R (2021) Reconstitution of muscle cell microtubule organization in vitro. Cytoskeleton (Hoboken) 78:492–502. https://doi.org/10.1002/cm.21710
107. Springstein B, Weissenbach J, Koch R, Stücker F, Stucken K (2020) The role of the cytoskeletal proteins MreB and FtsZ in multicellular cyanobacteria. FEBS Open Bio 10:2510–2531 [Internet]. https://doi.org/10.1002/2211-5463.13016. Cited 20 Oct 2021
108. Dyrka W, Coustou V, Daskalov A, Lends A, Bardin T, Berbon M et al (2020) Identification of NLR-associated amyloid signaling motifs in bacterial genomes. J Mol Biol 432:6005–6027 [Internet]. https://doi.org/10.1016/J.JMB.2020.10.004. Cited 20 Oct 2021
109. Volland J-M, Gonzalez-Rizzo S, Gros O, Tyml T, Ivanova N, Schulz F et al (2022) A centimeter-long bacterium with DNA compartmentalized in membrane-bound organelles. bioRxiv [Internet]. Cold Spring Harbor Laboratory 2022.02.16.480423. https://doi.org/10.1101/2022.02.16.480423. Cited 22 Apr 2022
110. Malan V, Vekemans M, Turleau C (2006) Chimera and other fertilization errors. Clin Genet 70:363–373 [Internet]. https://doi.org/10.1111/J.1399-0004.2006.00689.X. Cited 29 Sep 2021

111. Queremel Milani D, Chauhan P (2024) Genetics, Mosaicism [Internet]. StatPearls Publishing, Treasure Island (FL). https://www.ncbi.nlm.nih.gov/books/NBK559193/
112. Goss EM, Tabima JF, Cooke DEL, Restrepo S, Fry WE, Forbes GA et al (2014) The Irish potato famine pathogen *Phytophthora infestans* originated in central Mexico rather than the Andes. Proc Natl Acad Sci 111:8791–8796 [Internet]. https://doi.org/10.1073/pnas.1401884111. Cited 25 Sep 2020
113. Crouch JH, Crouch HK, Tenkouano A, Ortiz R (1999) VNTR-based diversity analysis of 2x and 4x full-sib Musa hybrids [Internet]. EJB Electron J Biotechnol. http://www.ejb.org/content/vol2/issue3/full/1/
114. Shakweer WME, Krivoruchko AY, Dessouki SM, Khattab AA (2023) A review of transgenic animal techniques and their applications. J Genet Eng Biotechnol 21:55. https://doi.org/10.1186/s43141-023-00502-z
115. Kumari R, Palaniyandi S, Hildebrandt GC (2018) Microbiome: an emerging new frontier in graft-versus-host disease. Dig Dis Sci [Internet]. Springer US. https://doi.org/10.1007/s10620-018-5369-9
116. Bordenstein SR, Theis KR (2015) Host biology in light of the microbiome: ten principles of holobionts and hologenomes. PLoS Biol 13 [Internet]. Public Library of Science. https://doi.org/10.1371/JOURNAL.PBIO.1002226. Cited 23 Apr 2022
117. Song YY, Zeng RS, Xu JF, Li J, Shen X, Yihdego WG (2010) Interplant communication of tomato plants through underground common mycorrhizal networks. PLoS ONE 5. https://doi.org/10.1371/journal.pone.0013324
118. Aratani Y, Uemura T, Hagihara T, Matsui K, Toyota M (2023) Green leaf volatile sensory calcium transduction in Arabidopsis. Nat Commun 14:6236 [Internet]. https://doi.org/10.1038/s41467-023-41589-9
119. Del Stabile F, Marsili V, Forti L, Arru L (2022) Is there a role for sound in plants? Plants (Basel, Switzerland) 11. https://doi.org/10.3390/plants11182391
120. Dimijian GG (2000) Evolving together: the biology of symbiosis, part 1. Proc Bayl Univ Med Cent 13:217–226 [Internet]. Baylor University Medical Center. http://www.ncbi.nlm.nih.gov/pubmed/16389385. Accessed 15 Jul 2019
121. Rogers S, Matheson T, Despland E, Dodgson T, Burrows M, Simpson S (2003) Mechanosensory-induced behavioural gregarization in the desert locust *Schistocerca gregaria*. J Exp Biol 206:3991–4002 [Internet]. https://doi.org/10.1242/JEB.00648. Cited 29 Sep 2021
122. Sword G, Simpson S, El Hadi O, Wilps H (2000) Density-dependent aposematism in the desert locust. Proc Biol Sci 267:63–68 [Internet]. https://doi.org/10.1098/RSPB.2000.0967. Cited 29 Sep 2021
123. van Wolferen M, Orell A, Albers S-V (2018) Archaeal biofilm formation. Nat Rev Microbiol 16:699–713 [Internet]. https://doi.org/10.1038/s41579-018-0058-4. Cited 8 Jul 2019
124. Sen CK, Mathew-Steiner SS, Das A, Sundaresan VB, Roy S (2020) Electroceutical management of bacterial biofilms and surgical infection. Antioxid Redox Signal 33:713–724. https://doi.org/10.1089/ars.2020.8086
125. Wang H, Tampio AJF, Xu Y, Nicholas BD, Ren D (2020) Noninvasive control of bacterial biofilms by wireless electrostimulation. ACS Biomater Sci Eng 6:727–738. https://doi.org/10.1021/acsbiomaterials.9b01199
126. Adam B, Baillie GS, Douglas LJ (2002) Mixed species biofilms of *Candida albicans* and *Staphylococcus epidermidis*. J Med Microbiol 51:344–349 [Internet]. https://doi.org/10.1099/0022-1317-51-4-344. Cited 18 Feb 2019
127. Abriat C, Virgilio N, Heuzey M-C, Daigle F (2019). Microbiological and real-time mechanical analysis of *Bacillus licheniformis* and *Pseudomonas fluorescens* dual-species biofilm. Microbiology [Internet]. https://doi.org/10.1099/mic.0.000819. Cited 10 Jun 2019
128. Babikova Z, Gilbert L, Bruce TJA, Birkett M, Caulfield JC, Woodcock C et al (2013) Underground signals carried through common mycelial networks warn neighbouring plants of aphid attack. Ecol Lett 16:835–843. https://doi.org/10.1111/ele.12115

129. Watts HJ, Veacute Ry A-A, Perera THS, Davies JM, Gow NAR (1998) Thigmotropism and stretch-activated channels in the pathogenic fungus *Candida albicans*. Microbiology 144(Pt 3):689–695. https://doi.org/10.1099/00221287-144-3-689
130. Bawa AS, Anilakumar KR (2013) Genetically modified foods: safety, risks and public concerns-a review. J Food Sci Technol 50:1035–1046 [Internet]. https://doi.org/10.1007/S13197-012-0899-1. Cited 23 Aug 2022
131. Shekhter AB (1986) Connective tissue as an integral system: role of cell-cell and cell-matrix interactions. Connect Tissue Res 15:23–31. https://doi.org/10.3109/03008208609001970
132. Sage EH (2001) Regulation of interactions between cells and extracellular matrix: a command performance on several stages. J Clin Invest 107:781–783. https://doi.org/10.1172/JCI12683
133. dos Santos CF, Acosta AL, Dorneles AL, dos Santos PDS, Blochtein B (2016) Queens become workers: pesticides alter caste differentiation in bees. Sci Rep 61:1–9 [Internet]. Nature Publishing Group. https://doi.org/10.1038/srep31605. Cited 29 Sep 2021
134. Kongsilp P, Taetragool U, Duangphakdee O (2024) Individual honey bee tracking in a beehive environment using deep learning and Kalman filter. Sci Rep 14:1061. https://doi.org/10.1038/s41598-023-44718-y
135. Zahadat P, Hahshold S, Thenius R, Crailsheim K, Thomas Schmickl T (2015) From honeybees to robots and back: division of labour based on partitioning social inhibition. Bioinspir Biomim 10:066005. https://doi.org/10.1088/1748-3190/10/6/066005. Cited 10 May 2019
136. Abdullaev K (2017) Funerary tradition of the ancient east in examples from anatolia and bactria-margiana: origins or parallels? TÜRKİYE BİLİMLER Akad ARKEOLOJİ DERGİSİ 20:27–49. https://doi.org/10.22520/tubaar.2017.20.002
137. Battin M (2009) Altruistic suicide. In: Bryant C, Peck D (eds) Encyclopedia of death and human experience (vol 1). SAGE Publications, Thousand Oaks/London/New Delhi/Singapore, pp 32–35
138. Maxwell R, Silverman P, Maxwell E (1983) The motive of geronticide. In: Sokolovsky J (ed) Aging aged third world. College of William and Mary, Williamsburg, pp 67–84
139. Young R (2013) Do elephant graveyards exist? [Internet]. Slate. https://slate.com/human-interest/2013/11/elephant-graveyards-do-they-exist.html
140. Azimzadeh J (2014) Exploring the evolutionary history of centrosomes. Philos Trans R Soc London Ser B, Biol Sci 369. https://doi.org/10.1098/rstb.2013.0453
141. Yamashita YM (2009) The centrosome and asymmetric cell division. Prion 3:84–88. https://doi.org/10.4161/pri.3.2.8821
142. Schatten G (1994) The centrosome and its mode of inheritance: the reduction of the centrosome during gametogenesis and its restoration during fertilization. Dev Biol 165:299–335 [Internet]. https://doi.org/10.1006/dbio.1994.1256
143. Goutas A, Trachana V (2021) Stem cells' centrosomes: how can organelles identified 130 years ago contribute to the future of regenerative medicine? World J Stem Cells 13:1177–1196. https://doi.org/10.4252/wjsc.v13.i9.1177
144. Pan Q, Guiguen Y, Herpin A (2018) Evolution of sex determining genes in fish. In: Skinner MKBT-E of R (Second E, editor). Academic Press, Oxford, pp 168–175. https://doi.org/10.1016/B978-0-12-809633-8.20552-9
145. Pla S, Benvenuto C, Capellini I, Piferrer F (2022) Switches, stability and reversals in the evolutionary history of sexual systems in fish. Nat Commun 13:3029. https://doi.org/10.1038/s41467-022-30419-z
146. Veyrunes F, Waters PD, Miethke P, Rens W, McMillan D, Alsop AE et al (2008) Bird-like sex chromosomes of platypus imply recent origin of mammal sex chromosomes. Genome Res 18:965–973. https://doi.org/10.1101/gr.7101908
147. DiCenzo G, Finan T (2017) The divided bacterial genome: structure, function, and evolution. Microbiol Mol Biol Rev 81:e00019-17. https://doi.org/10.1128/MMBR.00019-17
148. Lowen AC (2018) It's in the mix: reassortment of segmented viral genomes. PLoS Pathog 14:e1007200. https://doi.org/10.1371/journal.ppat.1007200
149. Sicard A, Michalakis Y, Gutié Rrez S, Phane Blanc S (2016) The strange lifestyle of multipartite viruses. https://doi.org/10.1371/journal.ppat.1005819

150. Kieft K, Anantharaman K (2022) Virus genomics: what is being overlooked? Curr Opin Virol 53:101200 [Internet]. https://doi.org/10.1016/j.coviro.2022.101200
151. Toussaint A, Chandler M (2012) Prokaryote genome fluidity: toward a system approach of the mobilome. Methods Mol Biol 57–80 [Internet]. https://doi.org/10.1007/978-1-61779-361-5_4. Cited 15 Jul 2019
152. Stokes HW, Gillings MR (2011) Gene flow, mobile genetic elements and the recruitment of antibiotic resistance genes into gram-negative pathogens [Internet]. FEMS Microbiol Rev 790–819. Oxford Academic. https://doi.org/10.1111/j.1574-6976.2011.00273.x. Cited 23 Sep 2020
153. Danchin EGJ (2016) Lateral gene transfer in eukaryotes: tip of the iceberg or of the ice cube. BMC Biol 14:1–3 [Internet]. BioMed Central Ltd. https://doi.org/10.1186/S12915-016-0330-X/FIGURES/1. Cited 23 Apr 2022
154. Nelson KE, Clayton RA, Gill SR, Gwinn ML, Dodson RJ, Haft DH et al (1999) Evidence for lateral gene transfer between archaea and bacteria from genome sequence of *Thermotoga maritima*. Nature 399:323–329 [Internet]. Nature Publishing Group. https://doi.org/10.1038/20601. Cited 4 Jul 2019
155. Taddeo S (2014) Intraspecies chimeras produced in laboratory settings (1960–1975) [Internet]. Embryo Project Encyclopedia, Arizona State University. https://hdl.handle.net/10776/8265
156. Rowland I, Gibson G, Heinken A, Scott K, Swann J, Thiele I et al (2018) Gut microbiota functions: metabolism of nutrients and other food components. Eur J Nutr 57:1–24 [Internet]. https://doi.org/10.1007/s00394-017-1445-8. Cited 11 Feb 2019
157. Gill SR, Pop M, Deboy RT, Eckburg PB, Turnbaugh PJ, Samuel BS et al (2006) Metagenomic analysis of the human distal gut microbiome. Science 312:1355–1359 [Internet]. https://doi.org/10.1126/science.1124234. Cited 20 Jun 2019
158. Sintim HO, Gürsoy UK (2016) Biofilms as "connectors" for oral and systems medicine: a new opportunity for biomarkers, molecular targets, and bacterial eradication. Omi A J Integr Biol 20:3–11 [Internet]. Mary Ann Liebert Inc. https://doi.org/10.1089/omi.2015.0146. Cited 29 Sep 2020
159. Dunny G, Brickman T, Dworkin M (2008) Multicellular behavior in bacteria: communication, cooperation, competition and cheating. Bioessays 30:296–298 [Internet]. https://doi.org/10.1002/BIES.20740. Cited 29 Sep 2021
160. Sikdar R, Elias M (2020) Quorum quenching enzymes and their effects on virulence, biofilm, and microbiomes: a review of recent advances. Expert Rev Anti Infect Ther 18:1221–1233. https://doi.org/10.1080/14787210.2020.1794815
161. Bhardwaj AK, Vinothkumar K, Rajpara N (2013) Bacterial quorum sensing inhibitors: attractive alternatives for control of infectious pathogens showing multiple drug resistance. Recent Pat Antiinfect Drug Discov 8:68–83. https://doi.org/10.2174/1574891x11308010012
162. Zhong S, He S (2021) Quorum sensing inhibition or quenching in *Acinetobacter baumannii*: the novel therapeutic strategies for new drug development. Front Microbiol 12 [Internet]. https://doi.org/10.3389/fmicb.2021.558003
163. Bakas S, Kambouris M (2021) The battle mechanics of the Hoplite Phalanx. Nuova Antol Mil 2021:2
164. Roth G (2015) Convergent evolution of complex brains and high intelligence. Philos Trans R Soc Lond Ser B, Biol Sci 370. https://doi.org/10.1098/rstb.2015.0049
165. Köhl J, Kolnaar R, Ravensberg WJ (2019) Mode of action of microbial biological control agents against plant diseases: relevance beyond efficacy. Front Plant Sci 10:845. Frontiers Media S.A. https://doi.org/10.3389/FPLS.2019.00845/BIBTEX
166. Nishida H (2020) Factors that affect the enlargement of bacterial protoplasts and spheroplasts. Int J Mol Sci 21. https://doi.org/10.3390/ijms21197131
167. Stingl K, Koraimann G (2017) Prokaryotic information games: how and when to take up and secrete DNA. Curr Top Microbiol Immunol 61–92 [Internet]. https://doi.org/10.1007/978-3-319-75241-9_3. Cited 15 Jul 2019
168. Dunning LT, Olofsson JK, Parisod C, Choudhury RR, Moreno-Villena JJ, Yang Y et al (2019) Lateral transfers of large DNA fragments spread functional genes among grasses. Proc Natl

Acad Sci 116:4416–4425 [Internet]. National Academy of Sciences. https://doi.org/10.1073/PNAS.1810031116. Cited 29 Jul 2020
169. Hu B, Huang S, Yin L (2021) The cytokine storm and COVID-19. J Med Virol 93:250–256 [Internet]. https://doi.org/10.1002/JMV.26232. Cited 20 Aug 2022
170. Fan J, Toth I, Stephenson RJ (2024) Recent scientific advancements towards a vaccine against Group A Streptococcus. Vaccines 12. https://doi.org/10.3390/vaccines12030272
171. Jeong SH, Shekhar N, Mutyala N, Canaday O (2024) Distinguishing acute rheumatic fever from post-streptococcal reactive arthritis. Cureus e55739. https://doi.org/10.7759/cureus.55739
172. Becker DE (2011) Adverse drug interactions. Anesth Prog 58:31–41
173. Wolf M (2015) Is there really such a thing as "one health"? Thinking about a more than human world from the perspective of cultural anthropology. Soc Sci Med 129:5–11. https://doi.org/10.1016/J.SOCSCIMED.2014.06.018
174. van Bruggen AHC, Goss EM, Havelaar A, van Diepeningen AD, Finckh MR, Morris JG (2019) One health—cycling of diverse microbial communities as a connecting force for soil, plant, animal, human and ecosystem health. Sci Total Environ 927–937. https://doi.org/10.1016/j.scitotenv.2019.02.091
175. Özdemir V (2020) "One Nature": a new vocabulary and frame for governance innovation in post-COVID-19 planetary health. OMICS 24:645–648 [Internet]. https://doi.org/10.1089/OMI.2020.0169. Cited 6 Sep 2021
176. Brodeur BR, Johnson WM, Diena BB, Visentin LP, Johnson KG (1977) Mitogenic activity of Neisseria gonorrhoeae surface antigens in mouse splenic lymphocyte culture. Can J Microbiol 23:1154–1160. https://doi.org/10.1139/m77-173
177. Stetzenbach LD (2009) Airborne infectious microorganisms. In: Schaechter MBT-E of M (Third E, editor). Academic Press, Oxford, pp 175–182. https://doi.org/10.1016/B978-012373944-5.00177-2
178. Carlson JA, Dabiri G, Cribier B, Sell S (2011) The immunopathobiology of syphilis: the manifestations and course of syphilis are determined by the level of delayed-type hypersensitivity. Am J Dermatopathol 33:433–460. https://doi.org/10.1097/DAD.0b013e3181e8b587
179. Kambouris ME, Manoussopoulos Y, Patrinos GP, Velegraki A (2019) Microbial genomics in public health: a translational risk-response aspect. In: Patrinos G (ed) Applied genomics and public health. Elsevier Academic Press, pp 131–148. https://doi.org/10.1016/B978-0-12-813695-9.00007-8
180. Lin A, Torres CM, Hobbs EC, Bardhan J, Aley SB, Spencer CT et al (2024) Computational and systems biology advances to enable bioagent agnostic signatures. Health Secur 22:130–139 [Internet]. Mary Ann Liebert, Inc. https://doi.org/10.1089/hs.2023.0076
181. Eisenkraft A (2024) Threat agnostic approach to epidemic management using continuous remote patient monitoring. Health Secur 22:156–158 [Internet]. Mary Ann Liebert, Inc. https://doi.org/10.1089/hs.2023.0061
182. Rubin E, Harvey C, Villatoro A, Dean B (2024) Next generation public health emergency readiness: standardized tools and a threat agnostic biosurveillance system. Health Secur 22:140–145 [Internet]. Mary Ann Liebert, Inc. https://doi.org/10.1089/hs.2023.0102
183. Leiser OP, Hobbs EC, Sims AC, Korch GW, Taylor KL (2021) Beyond the list: bioagent-agnostic signatures could enable a more flexible and resilient biodefense posture than an approach based on priority agent lists alone. Pathogens 10:1497 [Internet]. Multidisciplinary Digital Publishing Institute. https://doi.org/10.3390/PATHOGENS10111497. Cited 6 Oct 2022
184. Acevedo-Rocha CG, Budisa N (2016) Xenomicrobiology: a roadmap for genetic code engineering. Microb Biotechnol 9:666–676 [Internet]. Wiley. https://doi.org/10.1111/1751-7915.12398. Cited 21 Sep 2020
185. Schmidt M, Pei L, Budisa N (2018) Xenobiology: state-of-the-art, ethics, and philosophy of new-to-nature organisms. Adv Biochem Eng Biotechnol 162:301–316. https://doi.org/10.1007/10_2016_14

186. Kambouris ME, Gaitanis G, Manoussopoulos Y, Arabatzis M, Kantzanou M, Kostis GD et al (2018) Humanome versus microbiome: games of dominance and pan-biosurveillance in the Omics Universe. Omi A J Integr Biol. 22:528–538. https://doi.org/10.1089/omi.2018.0096
187. Kambouris ME, Manoussopoulos Y, Velegraki A (2021) The emerging pathogens: nature, status, and threat. In: Genomics in biosecurity, 1st edn. Elsevier Academic Press, London, pp 61–74
188. Woolhouse MEJ (2002). Population biology of emerging and re-emerging pathogens. Trends Microbiol s3–s7. Elsevier. https://doi.org/10.1016/S0966-842X(02)02428-9
189. Urbanski AH, Araujo JD, Creighton R, Nakaya HI (2019) Integrative biology approaches applied to human diseases. In: Husi H (ed) Computational biology (Internet). Brisbane (AU). https://doi.org/10.15586/computationalbiology.2019.ch2
190. Clarke R (2023) Introduction: cancer systems and integrative biology. Methods Mol Biol 2660:1–11. https://doi.org/10.1007/978-1-0716-3163-8_1
191. Siamoglou S, Boltsis I, Chassomeris CA, Kambouris ME (2020) Electromagnetism and the microbiome(s). In: Kambouris ME, Velegraki A (eds) Microbiomics: dimensions, applications, and translational implications of human and environmental microbiome research, 1st edn. Elsevier Academic Press, London, pp 299–331
192. Kambouris ME, Markogiannakis A, Arabatzis M, Manoussopoulos Y, Kantzanou M, Velegraki A (2017) Wireless electrostimulation: a new approach in combating infection? Future Microbiol 12:255–265. https://doi.org/10.2217/fmb-2017-0157
193. Kambouris M, Zagoriti Z, Lagoumintzis G, Poulas K (2014) From therapeutic electrotherapy to electroceuticals: formats, applications and prospects of electrostimulation. Annu Res Rev Biol 4:3054–3070. https://doi.org/10.9734/arrb/2014/10563
194. Cech TR, Steitz JA (2014) The noncoding RNA revolution-trashing old rules to forge new ones. Cell 157:77–94 [Internet]. https://doi.org/10.1016/j.cell.2014.03.008. Cited 11 Apr 2024
195. Neu TR, Lawrence JR (2014) Investigation of microbial biofilm structure by laser scanning microscopy. Adv Biochem Eng Biotechnol 1–51 [Internet]. https://doi.org/10.1007/10_2014_272. Cited 3 Jun 2019
196. Lythgoe MP, Mullish BH, Frampton AE, Krell J (2022) Polymorphic microbes: a new emerging hallmark of cancer. Trends Microbiol 30:1131–1134 [Internet]. https://doi.org/10.1016/j.tim.2022.08.004
197. Tang WHW, Li DY, Hazen SL (2019) Dietary metabolism, the gut microbiome, and heart failure. Nat Rev Cardiol 16:137–154 [Internet]. https://doi.org/10.1038/s41569-018-0108-7. Cited 5 Jul 2019
198. Tanaka M, Itoh H (2019) Hypertension as a metabolic disorder and the novel role of the gut. Curr Hypertens Rep 21:63 [Internet]. https://doi.org/10.1007/s11906-019-0964-5. Cited 27 Jun 2019
199. Enaud R, Vandenborght L-E, Coron N, Bazin T, Prevel R, Schaeverbeke T et al (2018) The mycobiome: a neglected component in the microbiota-gut-brain axis. Microorganisms 6:22 [Internet]. https://doi.org/10.3390/microorganisms6010022
200. Johnson KV-A, Foster KR (2018) Why does the microbiome affect behaviour? Nat Rev Microbiol 16:647–655 [Internet]. https://doi.org/10.1038/s41579-018-0014-3. Cited 5 Jul 2019
201. Lee CY, Abizaid A (2014) The gut–brain-axis as a target to treat stress-induced obesity. Front Endocrinol (Lausanne) 5 [Internet]. Frontiers Media SA. https://doi.org/10.3389/FENDO.2014.00117. Cited 20 Aug 2022
202. Ackermann J, Qu Q, LeGoff L, Ben Amar M (2022) Modeling the mechanics of growing epithelia with a bilayer plate theory. The European Physical Journal Plus 137(1). 10.1140/epjp/s13360-021-02205-1

MIX
Papier aus verantwortungsvollen Quellen
Paper from responsible sources
FSC® C105338
www.fsc.org

If you have any concerns about our products,
you can contact us on
ProductSafety@springernature.com

In case Publisher is established outside the EU,
the EU authorized representative is:
Springer Nature Customer Service Center GmbH
Europaplatz 3, 69115 Heidelberg, Germany

Printed by Libri Plureos GmbH
in Hamburg, Germany